Praise for the Hardcover Edition of Simon's
Probability Distributions Involving Gaussian Random Variables

In my research in wireless communications, I have found that I can get all the mathematical forms and material that I need from one source, whereas before I had to consult five or six references to get the complete information. There are a small number of reference works that have proven so invaluable that I have purchased a home copy in addition to my office copy. This handbook is one of them.
   -- Dr. Norman C. Beaulieu, University of Alberta

Our research Group at the University of Canterbury is engaged in both theoretical and experimental work on advanced wireless communications systems. We have found that the book by Simon, *Probability Distributions Involving Gaussian Random Variables* is an indispensable and often used reference for this work, so much so that we have purchased copies for the bookshelf in our laboratory. We do significant amounts of analysis involving Gaussian statistics and quadratic forms. The book is the one source that brings almost all of the necessary background material together in a readily usable form and is thus an extremely valuable reference.
   -- Desmond P. Taylor, University of Canterbury

The Gaussian distribution and those derived from it are at the very core of a huge number of problems in multi-disciplinary fields of engineering, mathematics and science. All scientists and professional engineers face, sooner or later, a problem whose solution involves processing, either analytically, or by simulation, or by a mixture of the two, random variables whose distribution is related to Gaussian variables. The book, with its comprehensive information in analytical, tabular, and graphical form, is an invaluable tool for scientists and engineers.
   -- Sergio Benedetto, Politecnico di Torino

T0214572

Praise for the Hardcover Edition of Simon's
Probability Distributions Involving Gaussian Random Variables

Because the source of randomness and noise is usually modeled as Gaussian, engineers and scientists often need the probability distributions of functions of Gaussian random variables. Marvin Simon has written an extremely useful handbook for everyone working in these fields. It is by far the most complete and easy-to-use collection of these characterizations of functions of Gaussian random variables.

Marvin Simon has created an important and very popular book that is useful to a very large community of engineers and scientists. This work is also very carefully done with many corrections to earlier works that were not easily available to this community.
    -- Jim Omura, Gordon and Betty Moore Foundation

Probability Distributions Involving Gaussian Random Variables by Marvin Simon is a unique and invaluable reference for researchers in communications, signal processing, math, statistics, and many other scientific fields. I have used this book in my own research and also recommended it to colleagues in universities and in industry, as well as to graduate students.
    -- Andrea Goldsmith, Stanford University

The reference book *Probability Distributions Involving Gaussian Random Variables*, authored by Dr. Marvin Simon, has become, in a very short time frame, one of the most useful aids to research in the field of digital communications that has come out in many years. It has numerous results that can save researchers in the field endless hours of work. It has replaced various other well known resources because of its concentration of relevant, timely, and easily accessible results.
    -- Larry Milstein, UCSD

PROBABILITY DISTRIBUTIONS INVOLVING GAUSSIAN RANDOM VARIABLES
A Handbook for Engineers and Scientists

THE SPRINGER INTERNATIONAL SERIES IN ENGINEERING AND COMPUTER SCIENCE

PROBABILITY DISTRIBUTIONS INVOLVING GAUSSIAN RANDOM VARIABLES
A Handbook for Engineers and Scientists

Marvin K. Simon
Principal Scientist
Jet Propulsion Laboratory
Pasadena, California, U.S.A.

 Springer

Library of Congress Control Number: 2006933933

Simon, Marvin K.
 Probability Distributions Involving Gaussian Random Variables
A Handbook for Engineers and Scientists/ by Marvin K. Simon

     p. cm.

ISBN: 978-1-4020-7058-7 (HC)   ISBN: 978-0-387-34657-1 (SC) E- ISBN: 978-0-387-47694-0

Printed on acid-free paper

First Softcover Printing, 2006

9 8 7 6 5 4 3 2 1

springer.com

This book is dedicated to my wife Anita, daughter Brette, and son Jeffrey who, through their devotion and dedication, allowed me to continuously burn the midnight oil. A tedious project such as this could have never come to fruition without the support, understanding, and encouragement they provided me all during the preparation of the manuscript.

CONTENTS

4. DIFFERENCE OF CHI-SQUARE RANDOM VARIABLES   25

5. SUM OF CHI-SQUARE RANDOM VARIABLES   35

6. PRODUCTS OF RANDOM VARIABLES   49

Johann Carl Friedrich Gauss
1777-1855

At the age of seven, Carl Friedrich Gauss started elementary school, and his potential was noticed almost immediately. His teacher, Büttner, and his assistant, Martin Bartels, were amazed when Gauss summed the integers from 1 to 100 instantly by spotting that the sum was 50 pairs of numbers each pair summing to 101.

In 1788 Gauss began his education at the Gymnasium with the help of Büttner and Bartels, where he learnt High German and Latin. After receiving a stipend from the Duke of Brunswick- Wolfenbüttel, Gauss entered Brunswick Collegium Carolinum in 1792. At the academy Gauss independently discovered Bode's law, the binomial theorem and the arithmetic- geometric mean, as well as the law of quadratic reciprocity and the prime number theorem.

In 1795 Gauss left Brunswick to study at Göttingen University. Gauss's teacher there was Kaestner, whom Gauss often ridiculed. His only known friend amongst the students was Farkas Bolyai. They met in 1799 and corresponded with each other for many years.

Gauss left Göttingen in - the construction of1798 without a diploma, but by this time he had made one of his most important discoveries a regular 17-gon by ruler and compasses This was the most major advance in this field since the time of Greek mathematics and was published as Section VII of Gauss's famous work, *Disquisitiones Arithmeticae.*

Gauss returned to Brunswick where he received a degree in 1799. After the Duke of Brunswick had agreed to continue Gauss's stipend, he requested that Gauss submit a doctoral dissertation to the University of Helmstedt. He already knew Pfaff, who was chosen to be his advisor. Gauss's dissertation was a discussion of the fundamental theorem of algebra.

With his stipend to support him, Gauss did not need to find a job so he devoted himself to research. He published the book *Disquisitiones Arithmeticae* in the summer of 1801. There were seven sections, all but the last section, referred to above, being devoted to number theory.

In June 1801, Zach, an astronomer whom Gauss had come to know two or three years previously, published the orbital positions of Ceres, a new "small planet" which was discovered by G Piazzi, an Italian astronomer on 1 January, 1801. Unfortunately, Piazzi had only been able to observe 9 degrees of its orbit before it disappeared behind the Sun. Zach published several predictions of its position, including one by Gauss which differed greatly from the others. When Ceres was rediscovered by Zach on 7 December 1801 it was almost exactly where Gauss had predicted. Although he did not disclose his methods at the time, Gauss had used his least squares approximation method.

In June 1802 Gauss visited Olbers who had discovered Pallas in March of that year and Gauss investigated its orbit. Olbers requested that Gauss be made director of the proposed new observatory in Göttingen, but no action was taken. Gauss began corresponding with Bessel, whom he did not meet until 1825, and with Sophie Germain.

Gauss married Johanna Ostoff on 9 October, 1805. Despite having a happy personal life for the first time, his benefactor, the Duke of Brunswick, was killed fighting for the Prussian army. In 1807 Gauss left Brunswick to take up the position of director of the Göttingen observatory.

Gauss arrived in Göttingen in late 1807. In 1808 his father died, and a year later Gauss's wife Johanna died after giving birth to their second son, who was to die soon after her. Gauss was shattered and wrote to Olbers asking him give him a home for a few weeks,

to gather new strength in the arms of your friendship – strength for a life which is only valuable because it belongs to my three small children.

Gauss was married for a second time the next year, to Minna the best friend of Johanna, and although they had three children, this marriage seemed to be one of convenience for Gauss.

Gauss's work never seemed to suffer from his personal tragedy. He published his second book, *Theoria motus corporum coelestium in sectionibus conicis Solem ambientium,* in 1809, a major two-volume treatise on the motion of celestial bodies. In the first volume he discussed differential equations, conic sections and elliptic orbits, while in the second volume, the main part of the work, he showed how to estimate and then to refine the estimation of a planet's orbit. Gauss's contributions to theoretical astronomy stopped after 1817, although he went on making observations until the age of 70.

Much of Gauss's time was spent on a new observatory, completed in 1816, but he still found the time to work on other subjects. His publications during this time include *Disquisitiones generales circa seriem infinitam,* a rigorous treatment of series and an introduction of the hypergeometric function, *Methodus nova integralium valores per approximationem inveniendi,* a practical essay on approximate integration, *Bestimmung der Genauigkeit der Beobachtungen,* a discussion of statistical estimators, and *Theoria attractionis corporum sphaeroidicorum ellipticorum homogeneorum methodus nova tractata.* The latter work was inspired by geodesic problems and was principally concerned with potential theory. In fact, Gauss found himself more and more interested in geodesy in the 1820's.

Gauss had been asked in 1818 to carry out a geodesic survey of the state of Hanover to link up with the existing Danish grid. Gauss was pleased to accept and took personal charge of the survey, making measurements during the day and reducing them at night, using his extraordinary mental capacity for calculations. He regularly wrote to Schumacher, Olbers and Bessel, reporting on his progress and discussing problems.

Because of the survey, Gauss invented the heliotrope which worked by reflecting the Sun's rays using a design of mirrors and a small telescope. However, inaccurate base lines were used for the survey and an unsatisfactory network of triangles. Gauss often wondered if he would have been better advised to have pursued some other occupation but he published over 70 papers between 1820 and 1830.

In 1822 Gauss won the Copenhagen University Prize with *Theoria attractionis...* together with the idea of mapping one surface onto another so that the two are similar in their smallest parts. This paper was published in 1825 and led to the much later publication of *Untersuchungen über Gegenstände der Höheren Geodäsie* (1843 and 1846). The paper *Theoria combinationis observationum erroribus minimis obnoxiae* (1823), with its supplement (1828), was devoted to mathematical statistics, in particular to the least squares method.

From the early 1800's Gauss had an interest in the question of the possible existence of a non-Euclidean geometry. He discussed this topic at length with Farkas Bolyai and in his correspondence with Gerling and Schumacher. In a book review in 1816 he discussed proofs which deduced the axiom of parallels from the other Euclidean axioms, suggesting that he believed in the existence of non-Euclidean geometry, although he was rather vague. Gauss confided in Schumacher, telling him that he believed his reputation would suffer if he admitted in public that he believed in the existence of such a geometry.

In 1831 Farkas Bolyai sent to Gauss his son János Bolyai's work on the subject. Gauss replied

to praise it would mean to praise myself .

Again, a decade later, when he was informed of Lobachevsky's work on the subject, he praised its "genuinely geometric" character, while in a letter to Schumacher in 1846, states that he

had the same convictions for 54 years

indicating that he had known of the existence of a non-Euclidean geometry since he was 15 years of age (this seems unlikely).

Gauss had a major interest in differential geometry, and published many papers on the subject. *Disquisitiones generales circa superficies curva* (1828) was his most renowned work in this field. In fact, this paper rose from his geodesic interests, but it contained such geometrical ideas as Gaussian curvature. The paper also includes Gauss's famous *theorema egregrium*:

If an area in E^3 can be developed (i.e. mapped isometrically) into another area of E^3, the values of the Gaussian curvatures are identical in corresponding points.

The period 1817-1832 was a particularly distressing time for Gauss. He took in his sick mother in 1817, who stayed until her death in 1839, while he was arguing with his wife and her family about whether they should go to Berlin. He had been offered a position at Berlin University and Minna and her family were keen to move there. Gauss, however, never liked change and decided to stay in Göttingen. In 1831 Gauss's second wife died after a long illness.

In 1831, Wilhelm Weber arrived in Göttingen as physics professor filling Tobias Mayer's chair. Gauss had known Weber since 1828 and supported his appointment. Gauss had worked on physics before 1831, publishing *Über ein neues allgemeines Grundgesetz der Mechanik*, which contained the principle of least constraint, and *Principia generalia theoriae figurae fluidorum in statu aequilibrii* which discussed forces of attraction. These papers were based on Gauss's potential theory, which proved of great importance in his work on physics. He later came to believe his potential theory and his method of least squares provided vital links between science and nature.

In 1832, Gauss and Weber began investigating the theory of terrestrial magnetism after Alexander von Humboldt attempted to obtain Gauss's assistance in making a grid of magnetic observation points around the Earth. Gauss was excited by this prospect and by 1840 he had written three important papers on the subject: *Intensitas vis magneticae terrestris ad mensuram absolutam revocata* (1832), *Allgemeine Theorie des Erdmagnetismus* (1839) and *Allgemeine Lehrsätze in Beziehung auf die im verkehrten Verhältnisse des Quadrats der Entfernung wirkenden Anziehungs- und Abstossungskräfte* (1840). These papers all dealt with the current theories on terrestrial magnetism, including Poisson's ideas, absolute measure for magnetic force and an empirical definition of terrestrial magnetism. Dirichlet's principle was mentioned without proof.

Allgemeine Theorie... showed that there can only be two poles in the globe and went on to prove an important theorem, which concerned the determination of the intensity of the horizontal component of the magnetic force along with the angle of inclination. Gauss used the Laplace equation to aid him with his calculations, and ended up specifying a location for the magnetic South pole.

Humboldt had devised a calendar for observations of magnetic declination. However, once Gauss's new magnetic observatory (completed in 1833 - free of all magnetic metals) had been built, he proceeded to alter many of Humboldt's procedures, not pleasing

Humboldt greatly. However, Gauss's changes obtained more accurate results with less effort.

Gauss and Weber achieved much in their six years together. They discovered Kirchhoff's laws, as well as building a primitive telegraph device which could send messages over a distance of 5000 ft. However, this was just an enjoyable pastime for Gauss. He was more interested in the task of establishing a world-wide net of magnetic observation points. This occupation produced many concrete results. The *Magnetischer Verein* and its journal were founded, and the atlas of geomagnetism was published, while Gauss and Weber's own journal in which their results were published ran from 1836 to 1841.

In 1837, Weber was forced to leave Göttingen when he became involved in a political dispute and, from this time, Gauss's activity gradually decreased. He still produced letters in response to fellow scientists' discoveries usually remarking that he had known the methods for years but had never felt the need to publish. Sometimes he seemed extremely pleased with advances made by other mathematicians, particularly that of Eisenstein and of Lobachevsky.

Gauss spent the years from 1845 to 1851 updating the Göttingen University widow's fund. This work gave him practical experience in financial matters, and he went on to make his fortune through shrewd investments in bonds issued by private companies.

Two of Gauss's last doctoral students were Moritz Cantor and Dedekind. Dedekind wrote a fine description of his supervisor

... usually he sat in a comfortable attitude, looking down, slightly stooped, with hands folded above his lap. He spoke quite freely, very clearly, simply and plainly: but when he wanted to emphasise a new viewpoint ... then he lifted his head, turned to one of those sitting next to him, and gazed at him with his beautiful, penetrating blue eyes during the emphatic speech. ... If he proceeded from an explanation of principles to the development of mathematical formulas, then he got up, and in a stately very upright posture he wrote on a blackboard beside him in his peculiarly beautiful handwriting: he always succeeded through economy and deliberate arrangement in making do with a rather small space. For numerical examples, on whose careful completion he placed special value, he brought along the requisite data on little slips of paper.

Gauss presented his golden jubilee lecture in 1849, fifty years after his diploma had been granted by Hemstedt University. It was appropriately a variation on his dissertation of 1799. From the mathematical community only Jacobi and Dirichlet were present, but Gauss received many messages and honours.

From 1850 onwards Gauss's work was again of nearly all of a practical nature although he did approve Riemann's doctoral thesis and heard his probationary lecture. His last known scientific exchange was with Gerling. He discussed a modified Foucalt pendulum in 1854. He was also able to attend the opening of the new railway link between Hanover and Göttingen, but this proved to be his last outing. His health deteriorated slowly, and Gauss died in his sleep early in the morning of 23 February, 1855.

Article by: J J O'Connor and E F Robertson
http://www-groups.dcs.st-and.ac.uk/~history/Mathematicians/Gauss.html

PREFACE

This book is intended for use by students, academicians and practicing engineers who in the course of their daily study or research have need for the probability distributions and associated statistics of random variables that are themselves Gaussian or in various forms derived from them. The format of the book is primarily that of a handbook in that, for the most part, the results are merely presented in their final form without derivation or discussion. As such the reader must rely on the typographical accuracy of the documented expressions, which the author has taken great pains to assure. Also included at the end of the book are numerous curves illustrating the behavior of a variety of the probability measures presented in mathematical form.

The author wishes to acknowledge his many colleagues in industry and academia for the encouragement and support they provided for this project without which it might never have gotten started.

INTRODUCTION

There are certain reference works that engineers and scientists alike find invaluable in their day-to-day work activities. Many of these reference volumes are of a generic nature such as tables of integrals, tables of series, handbooks of mathematical formulas and transforms, etc. (see Refs. 1, 2, 3, and 4 for example), whereas others are collections of technical papers and textbooks that directly relate to the individual's specific field of specialty. Continuing along this train of thought, there exists a great deal of valuable information that, in its original form, was published in university and company reports and as such the general public was in many cases not aware of its existence. Even worse, today this archival material is no longer available to the public in any form since its original source has declared it to be out of print for quite some time now. Furthermore, most of the authors of these works have long since retired or sadder yet have passed on; however, the material contained in the documents themselves has intrinsic value and is timeless in terms of its value to today's practicing engineer or scientist. As time marches on and new young engineers and scientists replace the old ones, the passing of the torch must include a means by which this valuable information be communicated to the new generation. Such is the primary motivation behind this book, the more detailed objective being described as follows.

One of the most important, from both the theoretical and practical viewpoint, probability distributions is the *Gaussian* distribution or as mathematicians prefer to call it the *normal* distribution. Although the statistical characterization of the basic Gaussian random variable (RV), e.g., its probability density function (PDF), cumulative distribution function (CDF), and characteristic function (CF) are well-known and widely documented in the literature (e.g., [5]), in dealing with the applications, one is quite often in need of similar characterizations for arithmetic combinations, e.g., sums, differences, products, ratios of Gaussian RVs and also the square of Gaussian RVs (so-called *chi-square* RVs.) Other applications involve *log-normal* RVs and thus their statistical characterization is also of interest. Still other applications involve the maximum or minimum of two or more RVs. Although many published references on the above subjects implicitly make use of such PDFs, CDFs, and CFs, a publicly available compilation of these probability measures for a wide variety of cases is difficult to find. Perhaps the most comprehensive source for a large number of such probability measures is a Stanford University report by J. Omura and T. Kailath [6] that dates back more than three and a half decades. To those who are aware of its existence, this report has become somewhat of a classic reference on the subject and considering the fact that it was prepared prior to the age of personal computers and desktop publishing, it should be looked upon as a remarkable feat. Because of this nature of preparation, however, its absolute reliability suffers from the fact that it contains crucial typographical as well as other miscellaneous errors. These errors were discovered by this author when trying to apply the results in Omura and Kailath [6] to his own research. Furthermore, although the report includes a list of references from which the results contained therein were either obtained or derived, many of these references are no longer available (e.g., unpublished reports from companies which, in some instances, no longer exist, books that are out of print, etc.). Thus, the ability to fix the incorrect results is indeed hampered by the lack of availability of the references.

With the above in mind, the author set out to find the correct results and document them here for the benefit of: (a) those who have Ref. 6 and have discovered the existence of the errors but didn't care to take the time to correct them, (b) those who have Ref. 6 and have not yet found or are unaware of the existence of the errors, or (c) those who have never seen the report in the first place and would benefit from having these results. In addition, for the results that are indeed correct in Ref. 6, we have, in many instances, expressed them in a form which has a more pleasing appearance and lends itself better to the applications at hand. Furthermore, although the book does not focus on any particular application, from time to time specific applications and the location of their discussion in the literature will be mentioned as evidence of the usefulness of the results contained herein.

From the standpoint of the results themselves, the goal is to express them, as far as possible, in terms of well-known (and tabulated) functions. (In some instances, this requires an expression in the form of a single (occasionally double) infinite series.) When this goal is not achievable, as is more typical for CDFs than for PDFs, a set of dashes will be used to indicate the lack of availability of such results. While in keeping with the style of Ref. 6 this book has the flavor of a reference handbook for researchers, it also serves as a valuable companion to a college textbook used to teach a course on probability and random processes.

1

BASIC DEFINITIONS AND NOTATION

Throughout the book, a variable with a boldface type will be used to denote a matrix or vector, the latter always assumed to be in column format. Random variables are denoted by upper case letters whereas the values that these variables assume are denoted by lower case letters. Finally, $p(\cdot)$ is used to denote a PDF, $P(\cdot)$ is used to denote a CDF, and $\Psi(\cdot)$ is used to denote a CF.

A real *Gaussian* random variable (RV) is defined as one having the probability density function (PDF)

$$p_X(x) = \frac{1}{\sqrt{2\pi\sigma_X^2}} \exp\left[-\frac{(x-\overline{X})^2}{2\sigma_X^2}\right] \tag{1.1}$$

where $\overline{X} \triangleq E\{X\}$ is the statistical mean of X and σ_X^2 is the variance of X. The corresponding CDF, $P_X(x)$, and CF, $\Psi_X(\omega)$, are given by

$$P_X(x) = 1 - Q\left(\frac{\overline{X}}{\sigma_X}\right) \tag{1.2}$$

where

$$Q(x) \triangleq \frac{1}{\sqrt{2\pi}} \int_x^\infty \exp\left(-\frac{1}{2}y^2\right) dy \tag{1.3}$$

is the Gaussian probability integral which is related to the complementary error function by

$$Q(x) = \frac{1}{2}\operatorname{erfc}\left(\frac{x}{\sqrt{2}}\right) \tag{1.4}$$

and

$$\Psi_X(\omega) = E\{e^{j\omega X}\} = \exp\left(j\omega\overline{X} - \frac{\omega^2 \sigma_X^2}{2} \right) \qquad (1.5)$$

with $j = \sqrt{-1}$. A complex Gaussian RV $\tilde{X} = X_R + jX_I$ is one in which X_R and X_I each have a PDF of the form in (1.1). The PDF of $\tilde{X}$ is represented by the joint PDF of X_R and X_I.

For the purpose of this book, a Gaussian vector **X** is a vector whose components are independent Gaussian RVs with equal variance σ_X^2 but in general different statistical means. That is, for

$$\mathbf{X} = \begin{bmatrix} X_1 \\ X_2 \\ . \\ . \\ X_n \end{bmatrix} \qquad (1.6)$$

we have

$$E\{\mathbf{X}\} = \overline{\mathbf{X}} = \begin{bmatrix} \overline{X}_1 \\ \overline{X}_2 \\ . \\ . \\ \overline{X}_n \end{bmatrix}, \quad \sigma_{X_i}^2 = \sigma^2, i = 1, 2, ..., n \qquad (1.7)$$

The class of Gaussian vectors described above shall be denoted by $N_n(\overline{\mathbf{X}}, \sigma^2)$. Thus, $\mathbf{X} \in N_n(\overline{\mathbf{X}}, \sigma^2)$ means that **X** is a vector having the properties in (1.7).

Only Gaussian vectors of the same dimension are allowed to be statistically independent. In particular, if $\mathbf{X}^{(1)} \in N_n(\overline{\mathbf{X}}^{(1)}, \sigma_1^2)$ and $\mathbf{X}^{(2)} \in N_n(\overline{\mathbf{X}}^{(2)}, \sigma_2^2)$ are dependent Gaussian vectors, then for

$$\mathbf{X}^{(1)} = \begin{bmatrix} X_1^{(1)} \\ X_2^{(1)} \\ . \\ . \\ X_n^{(1)} \end{bmatrix}, \mathbf{X}^{(2)} = \begin{bmatrix} X_1^{(2)} \\ X_2^{(2)} \\ . \\ . \\ X_n^{(2)} \end{bmatrix}, \overline{\mathbf{X}}^{(1)} = \begin{bmatrix} \overline{X}_1^{(1)} \\ \overline{X}_2^{(1)} \\ . \\ . \\ \overline{X}_n^{(1)} \end{bmatrix}, \overline{\mathbf{X}}^{(2)} = \begin{bmatrix} \overline{X}_1^{(2)} \\ \overline{X}_2^{(2)} \\ . \\ . \\ \overline{X}_n^{(2)} \end{bmatrix} \qquad (1.8)$$

we have

$$E\left\{\left(X_j^{(1)} - \overline{X}_j^{(1)}\right)\left(X_k^{(2)} - \overline{X}_k^{(2)}\right)\right\} = \begin{cases} 0, & j \neq k \\ \\ \rho\sigma_1\sigma_2, & j = k \end{cases} \qquad j,k = 1,2,...,n \qquad (1.9)$$

That is, only components of $\mathbf{X}^{(1)}$ and $\mathbf{X}^{(2)}$ having identical subscripts can be correlated. This dependence is summarized by the covariance matrix for each pair of components

$$\mathbf{M} = \begin{bmatrix} \sigma_1^2 & \rho\sigma_1\sigma_2 \\ \rho\sigma_1\sigma_2 & \sigma_2^2 \end{bmatrix} \qquad (1.10)$$

whose inverse is given by

$$\mathbf{M}^{-1} \triangleq \mathbf{W} = \begin{bmatrix} w_{11} & w_{12} \\ w_{21} & w_{22} \end{bmatrix} = \frac{1}{1-\rho^2} \begin{bmatrix} \dfrac{1}{\sigma_1^2} & -\dfrac{\rho}{\sigma_1\sigma_2} \\ -\dfrac{\rho}{\sigma_1\sigma_2} & \dfrac{1}{\sigma_2^2} \end{bmatrix} \qquad (1.11)$$

Clearly if $\mathbf{X}^{(1)}$ and $\mathbf{X}^{(2)}$ are independent, then $\rho = 0$.

A *Rayleigh* RV, R, of order n is defined by the norm of a zero mean Gaussian vector. That is, if $\mathbf{X} \in N_n(\mathbf{0},\sigma^2)$, then

$$R = \|\mathbf{X}\| = \sqrt{\sum_{k=1}^{n} X_k^2} \qquad (1.12)$$

A *Rician* RV, R, of order n is defined analogously by the norm of a nonzero mean Gaussian vector, $\mathbf{X} \in N_n(\overline{\mathbf{X}},\sigma^2)$.[1]

A *chi-square* RV is defined as the squared norm of a Gaussian vector. If the Gaussian vector is zero mean, then

---

[1] The more common usage of the terms Rayleigh RV and Rician RV refers to the specific case $n=2$. Here, for convenience, we use them in their extended form. We also note that the Rayleigh distribution of order m is equivalent to the *Nakamagi-m* distribution [7] when m is integer. Likewise the Rician distribution of order 2 is equivalent to the *Nakagami-n* distribution [7].

$$Y = \|\mathbf{X}\|^2 = \sum_{k=1}^{n} X_k^2 \tag{1.13}$$

is referred to as a *central chi-square* RV with n degrees of freedom. For $n = 1$, the central chi-square distribution simplifies to an *exponential* distribution. Furthermore, for $n \neq 1$, the central chi-square distribution is equivalent to the *gamma* distribution with parameter n when n is integer. If the Gaussian vector is nonzero mean, then Y of (1.13) is referred to as a *noncentral chi-square RV* with n degrees of freedom and *noncentrality parameter*

$$a_y^2 = \|\overline{\mathbf{X}}\|^2 = \sum_{k=1}^{n} \overline{X}_k^2 \tag{1.14}$$

Whereas the book will only provide results for the noncentral moments of a RV, e.g., $E\{X^k\}$, k integer, the central moments, most notably the variance, can be obtained from the relation

$$E\{(X - \overline{X})^k\} = \sum_{i=0}^{k} \binom{k}{i}(-1)^{k-i} \overline{X}^{k-i} E\{X^i\}, k \text{ integer} \tag{1.15}$$

where $\binom{k}{i} = \dfrac{k!}{i!(k-i)!}$ denotes a combinatorial coefficient.

A *log-normal* RV is one whose logarithm has a Gaussian distribution. That is $\gamma = 10^{X/10}$ is a log-normal RV when X is characterized by the PDF of (1.1).

2

FUNDAMENTAL ONE-DIMENSIONAL VARIABLES

A. Gaussian

The PDF, CDF, and CF of a Gaussian RV $\mathbf{X} \in N_n(\overline{\mathbf{X}}, \sigma^2)$ are given in (1.1), (1.2), and (1.5) respectively with σ_X replaced by σ. For $\overline{\mathbf{X}} = 0$, the even moments of the components of $\mathbf{X}$ are given by

$$E\{X_i^{2k}\} = \overline{X_i^{2k}} = \frac{(2k)!}{k!}\left(\frac{1}{2}\sigma^2\right)^k, \, k \text{ integer} \qquad (2.1)$$

All odd moments are equal to zero.

B. Rayleigh

1. $n = 1$

Since the square root always yields a positive quantity, then by definition $R = \sqrt{X^2} = |X|$, i.e., a single-sided Gaussian RV with PDF and CDF given by

$$p_R(r) = \frac{2}{\sqrt{2\pi\sigma^2}} \exp\left(-\frac{r^2}{2\sigma^2}\right), r \geq 0 \qquad (2.2)$$

$$P_R(r) = 1 - 2Q\left(\frac{r}{\sigma}\right), r \geq 0 \qquad (2.3)$$

Also, the moments of R are given by

$$E\{R^k\} = \frac{(2\sigma^2)^{k/2}}{\sqrt{\pi}} \Gamma\left(\frac{k+1}{2}\right), k \text{ integer} \tag{2.4}$$

2. $n = 2$

Here R corresponds to a conventional Rayleigh RV with PDF and CDF

$$p_R(r) = \frac{r}{\sigma^2} \exp\left(-\frac{r^2}{2\sigma^2}\right), r \geq 0 \tag{2.5}$$

$$P_R(r) = 1 - \exp\left(-\frac{r^2}{2\sigma^2}\right), r \geq 0 \tag{2.6}$$

Also, the moments of R are given by

$$E\{R^k\} = (2\sigma^2)^{k/2} \Gamma\left(1 + \frac{k}{2}\right), k \text{ integer} \tag{2.7}$$

3. $n = 2m$

$$p_R(r) = \frac{2r^{2m-1}}{(2\sigma^2)^m (m-1)!} \exp\left(-\frac{r^2}{2\sigma^2}\right), r \geq 0 \tag{2.8}$$

$$P_R(r) = 1 - \exp\left(-\frac{r^2}{2\sigma^2}\right) \sum_{i=0}^{m-1} \frac{1}{i!} \left(\frac{r^2}{2\sigma^2}\right)^i, r \geq 0 \tag{2.9}$$

$$E\{R^k\} = (2\sigma^2)^{k/2} \frac{\Gamma\left(m+\frac{k}{2}\right)}{(m-1)!}, k \text{ integer} \tag{2.10}$$

4. $n = 2m + 1$

$$p_R(r) = \frac{2r^{2m}}{(2\sigma^2)^{m+1/2} \Gamma(m+1/2)} \exp\left(-\frac{r^2}{2\sigma^2}\right), r \geq 0 \tag{2.11}$$

$$P_R(r) = ------- \tag{2.12}$$

$$E\{R^k\} = (2\sigma^2)^{k/2} \frac{\Gamma\left(m + \frac{k+1}{2}\right)}{\Gamma(m+1/2)}, k \text{ integer} \qquad (2.13)$$

C. Rician

1. $n = 1$

$$p_R(r) = \frac{1}{\sqrt{2\pi\sigma^2}} \exp\left(-\frac{r^2+a^2}{2\sigma^2}\right)\left[\exp\left(\frac{ar}{\sigma^2}\right) + \exp\left(-\frac{ar}{\sigma^2}\right)\right], r \geq 0 \quad (2.14)$$

$$P_R(r) = Q\left(\frac{a-r}{\sigma}\right) - Q\left(\frac{a+r}{\sigma}\right), r \geq 0 \qquad (2.15)$$

$$E\{R^k\} = (2\sigma^2)^{k/2} \exp\left(-\frac{a^2}{2\sigma^2}\right) \frac{\Gamma\left(\frac{k+1}{2}\right)}{\sqrt{\pi}} {}_1F_1\left(\frac{k+1}{2}; \frac{1}{2}; \frac{a^2}{2\sigma^2}\right), k \text{ integer} \quad (2.16)$$

where ${}_1F_1(\alpha; \beta; \gamma)$ is the confluent hypergeometric function [2] and $a = |\overline{X}|$.

2. $n = 2$

Here R corresponds to a conventional Rician RV with parameter $a = \|\overline{\mathbf{X}}\|$.

$$p_R(r) = \frac{r}{\sigma^2} \exp\left(-\frac{r^2+a^2}{2\sigma^2}\right) I_0\left(\frac{ra}{\sigma^2}\right), r \geq 0 \qquad (2.17)$$

$$P_R(r) = 1 - Q_1\left(\frac{a}{\sigma}, \frac{r}{\sigma}\right), r \geq 0 \qquad (2.18)$$

$$E\{R^k\} = (2\sigma^2)^{k/2} \exp\left(-\frac{a^2}{2\sigma^2}\right) \Gamma\left(1+\frac{k}{2}\right) {}_1F_1\left(1+\frac{k}{2}, 1; \frac{a^2}{2\sigma^2}\right), k \text{ integer}$$

$$(2.19)$$

where

$$Q_1(\alpha,\beta) = \int_\beta^\infty x \exp\left(-\frac{x^2+\alpha^2}{2}\right) I_0(\alpha x)\,dx \tag{2.20}$$

is the first-order Marcum Q-function [8].[2]

3. $n = 2m$

$$P_R(r) = \frac{r^m}{\sigma^2 a^{m-1}} \exp\left(-\frac{r^2+a^2}{2\sigma^2}\right) I_{m-1}\left(\frac{ra}{\sigma^2}\right),\ r \geq 0 \tag{2.21}$$

$$P_R(r) = 1 - Q_m\left(\frac{a}{\sigma},\frac{r}{\sigma}\right),\ r \geq 0 \tag{2.22}$$

$$E\{R^k\} = \left(2\sigma^2\right)^{k/2} \exp\left(-\frac{a^2}{2\sigma^2}\right) \frac{\Gamma\left(m+\frac{k}{2}\right)}{(m-1)!}\,{}_1F_1\left(m+\frac{k}{2};m;\frac{a^2}{2\sigma^2}\right),\ k\ \text{integer} \tag{2.23}$$

where

$$Q_m(\alpha,\beta) = \frac{1}{\alpha^{m-1}} \int_\beta^\infty x^m \exp\left(-\frac{x^2+\alpha^2}{2}\right) I_m(\alpha x)\,dx \tag{2.24}$$

is the generalized (mth-order) Marcum Q-function [8].

4. $n = 2m+1$

$$P_R(r) = \frac{1}{\sqrt{2\pi\sigma^2}} \left(\frac{r}{a}\right)^m \exp\left(-\frac{r^2+a^2}{2\sigma^2}\right) \left[\exp\left(\frac{ra}{\sigma^2}\right) \sum_{i=0}^{m-1} \frac{(-1)^i (m+i-1)!}{i!(m-i-1)!}\left(\frac{\sigma^2}{2ra}\right)^i\right.$$

$$\left. +(-1)^m \exp\left(-\frac{ra}{\sigma^2}\right) \sum_{i=0}^{m-1} \frac{(m+i-1)!}{i!(m-i-1)!}\left(\frac{\sigma^2}{2ra}\right)^i \right],\ r \geq 0 \tag{2.25}$$

$$P_R(r) = ------- \tag{2.26}$$

---

[2] More often than not in the literature, the subscript "1" identifying the order of the first-order Marcum Q-function is dropped from the notation. We shall maintain its identity in this text to avoid possible ambiguity with the two-dimensional Gaussian Q-function defined in Eq. (A.37) of Appendix A.

$$E\{R^k\} = (2\sigma^2)^{k/2} \exp\left(-\frac{a^2}{2\sigma^2}\right) \frac{\Gamma\left(m + \frac{k+1}{2}\right)}{\Gamma\left(m + \frac{1}{2}\right)} {}_1F_1\left(m + \frac{k+1}{2}; m + \frac{1}{2}; \frac{a^2}{2\sigma^2}\right), \quad (2.27)$$

$$k \text{ integer}$$

D. Central Chi-Square

1. $n = 1$

$$p_Y(y) = \frac{1}{\sqrt{2\pi\sigma^2 y}} \exp\left(-\frac{y}{2\sigma^2}\right), \quad y \geq 0 \qquad (2.28)$$

$$P_Y(y) = -------- \qquad (2.29)$$

$$\Psi_Y(\omega) = \left(\frac{1}{1 - 2j\omega\sigma^2}\right)^{1/2} \qquad (2.30)$$

$$E\{Y^k\} = (2\sigma^2)^k \frac{\Gamma(k + 1/2)}{\sqrt{\pi}}, \quad k \text{ integer} \qquad (2.31)$$

2. $n = 2m$

$$p_Y(y) = \frac{1}{2\sigma^2 \Gamma(m)} \left(\frac{y}{2\sigma^2}\right)^{m-1} \exp\left(-\frac{y}{2\sigma^2}\right), \quad y \geq 0 \qquad (2.32)$$

$$P_Y(y) = 1 - \exp\left(-\frac{y}{2\sigma^2}\right) \sum_{i=0}^{m-1} \frac{1}{i!} \left(\frac{y}{2\sigma^2}\right)^i, \quad y \geq 0 \qquad (2.33)$$

$$\Psi_Y(\omega) = \left(\frac{1}{1 - 2j\omega\sigma^2}\right)^m \qquad (2.34)$$

Since the kth moment of a central chi-square RV with $2m$ degrees of freedom is equal to the $2k$th moment of a Rayleigh RV of order $2m$, then it is straight-forward to obtain the moments of Y from (2.10) as

$$E\{Y^k\} = (2\sigma^2)^k \frac{\Gamma(m+k)}{(m-1)!}, \; k \text{ integer}$$ (2.35)

3. $n = 2m + 1$

$$p_Y(y) = \frac{1}{2\sigma^2 \Gamma(m+1/2)} \left(\frac{y}{2\sigma^2}\right)^{m-1/2} \exp\left(-\frac{y}{2\sigma^2}\right), \; y \geq 0$$ (2.36)

$$P_Y(y) = - - - - - -$$ (2.37)

$$\Psi_Y(s) = \left(\frac{1}{1 - 2js\sigma^2}\right)^{m+1/2}$$ (2.38)

$$E\{Y^k\} = (2\sigma^2)^k \frac{\Gamma\left(m+k+\dfrac{1}{2}\right)}{\Gamma\left(m+\dfrac{1}{2}\right)}, \; k \text{ integer}$$ (2.39)

E. Noncentral Chi-Square

1. $n = 1$

$$p_Y(y) = \frac{1}{2\sqrt{2\pi\sigma^2 y}} \exp\left(-\frac{y+a^2}{2\sigma^2}\right)\left[\exp\left(\sqrt{\frac{a^2 y}{\sigma^4}}\right) + \exp\left(-\sqrt{\frac{a^2 y}{\sigma^4}}\right)\right], \; y \geq 0$$ (2.40)

$$P_Y(y) = - - - - - -$$ (2.41)

$$\Psi_Y(\omega) = \left(\frac{1}{1 - 2j\omega\sigma^2}\right)^{1/2} \exp\left(\frac{j\omega a^2}{1 - 2j\omega\sigma^2}\right)$$ (2.42)

$$E\{Y^k\} = (2\sigma^2)^k \exp\left(-\frac{a^2}{2\sigma^2}\right)\frac{\Gamma(k+1/2)}{\sqrt{\pi}} {}_1F_1\left(k+\frac{1}{2};\frac{1}{2};\frac{a^2}{2\sigma^2}\right), \; k \text{ integer}$$ (2.43)

2. $n = 2m$

$$p_Y(y) = \frac{1}{2\sigma^2}\left(\frac{y}{a^2}\right)^{(m-1)/2} \exp\left(-\frac{y+a^2}{2\sigma^2}\right) I_{m-1}\left(\sqrt{\frac{a^2 y}{\sigma^4}}\right), y \geq 0 \qquad (2.44)$$

$$P_Y(y) = 1 - Q_m\left(\frac{a}{\sigma}, \frac{\sqrt{y}}{\sigma}\right), y \geq 0 \qquad (2.45)$$

$$\Psi_Y(\omega) = \left(\frac{1}{1-2j\omega\sigma^2}\right)^m \exp\left(\frac{j\omega a^2}{1-2j\omega\sigma^2}\right) \qquad (2.46)$$

Since the kth moment of a noncentral chi-square RV with $2m$ degrees of freedom is equal to the $2k$th moment of a Rician RV of order $2m$, then it is straightforward to obtain the moments of Y from (2.23) as

$$E\{Y^k\} = (2\sigma^2)^k \exp\left(-\frac{a^2}{2\sigma^2}\right)\frac{(m+k-1)!}{(m-1)!} {}_1F_1\left(m+k; m; \frac{a^2}{2\sigma^2}\right), k \text{ integer} \quad (2.47)$$

3. $n = 2m+1$

$$p_Y(y) = \frac{1}{2\sigma^2}\left(\frac{y}{a^2}\right)^{(m-1/2)/2} \exp\left(-\frac{y+a^2}{2\sigma^2}\right) I_{m-1/2}\left(\sqrt{\frac{a^2 y}{\sigma^4}}\right), y \geq 0 \qquad (2.48)$$

$$P_Y(y) = ------ \qquad (2.49)$$

$$\Psi_Y(\omega) = \left(\frac{1}{1-2j\omega\sigma^2}\right)^{m+1/2} \exp\left(\frac{j\omega a^2}{1-2j\omega\sigma^2}\right) \qquad (2.50)$$

$$E\{Y^k\} = (2\sigma^2)^k \exp\left(-\frac{a^2}{2\sigma^2}\right)\frac{\Gamma\left(m+k+\frac{1}{2}\right)}{\Gamma\left(m+\frac{1}{2}\right)} {}_1F_1\left(m+k+\frac{1}{2}; m+\frac{1}{2}; \frac{a^2}{2\sigma^2}\right), \qquad (2.51)$$

$$k \text{ integer}$$

F. Log-Normal

Let $X \in N_1(\overline{X}, \sigma^2)$. Then the PDF of $\gamma = 10^{X/10}$ is given by

$$p_\gamma(\gamma) = \frac{\xi}{\sqrt{2\pi}\sigma\gamma} \exp\left[-\frac{\left(10\log_{10}\gamma - \overline{X}\right)^2}{2\sigma^2}\right], \gamma \geq 0 \qquad (2.52)$$

$$P_\gamma(\gamma) = 1 - Q\left(\frac{10\log_{10}\gamma - \overline{X}}{\sigma}\right), \gamma \geq 0 \qquad (2.53)$$

where $\xi = 10/\ln 10$ and $\overline{X}$ (dB) and σ^2 (dB) correspond to the mean and variance of $10\log_{10}\gamma$. The CF of γ is not obtainable in closed form but can be approximated by a Gauss-Hermite expansion as

$$\Psi_\gamma(\omega) \cong \frac{1}{\sqrt{\pi}} \sum_{n=1}^{N_p} H_{x_n} \exp\left(10^{\left(\sqrt{2}\sigma x_n + \overline{X}\right)/10} j\omega\right) \qquad (2.54)$$

where x_n are the zeros and H_{x_n} are the weight factors of the N_p-order Hermite polynomial and can be found in Table 25.10 of [2]. In addition, the moments of γ are given by

$$E\{\gamma^k\} = \exp\left[\frac{k}{\xi}\overline{X} + \frac{1}{2}\left(\frac{k}{\xi}\right)^2\sigma^2\right], k \text{ integer} \qquad (2.55)$$

3

FUNDAMENTAL MULTIDIMENSIONAL VARIABLES

A. Gaussian

The joint PDF of a Gaussian vector $\mathbf{X} \in N_n(\overline{\mathbf{X}}, \sigma^2)$ with covariance matrix

$$\mathbf{M}_X = E\left\{(\mathbf{X} - \overline{\mathbf{X}})(\mathbf{X} - \overline{\mathbf{X}})'\right\} \tag{3.1}$$

is given as

$$p_X(\mathbf{x}) = \frac{1}{(2\pi)^{n/2}|\mathbf{M}_X|^{1/2}} \exp\left[-\frac{1}{2}(\mathbf{x} - \overline{\mathbf{X}})'\mathbf{M}_X^{-1}(\mathbf{x} - \overline{\mathbf{X}})\right] \tag{3.2}$$

where $|\mathbf{M}_X|$ denotes the determinant of $\mathbf{M}_X$. Letting

$$\boldsymbol{\omega} = \begin{bmatrix} \omega_1 \\ \omega_1 \\ . \\ . \\ \omega_n \end{bmatrix} \tag{3.3}$$

then the joint CF is given by

$$\Psi_X(\boldsymbol{\omega}) = \exp\left[j\boldsymbol{\omega}'\overline{\mathbf{X}} - \frac{1}{2}\boldsymbol{\omega}'\mathbf{M}_X\boldsymbol{\omega}\right] \tag{3.4}$$

For the special case of $n = 2, \overline{\mathbf{X}} = 0$, and the covariance matrix of (1.10), the joint moments are given as

$$E\{X_1^{k_1} X_2^{k_2}\} = \begin{cases} 0, & k_1 + k_2 \text{ odd} \\ \sigma^{k_1+k_2} \sum_{i=0}^{\lfloor k_1/2 \rfloor} \binom{k_1}{2i}(1-\rho^2)^i \rho^{k_1-2i}(k_1+k_2-2i-1)!!(2i-1)!!, \\ & k_1 + k_2 \text{ even} \end{cases} \qquad (3.5)$$

B. Rayleigh

Consider the pair of Rayleigh RVs of order n, $R_1 = \|\mathbf{X}^{(1)}\|$, $R_2 = \|\mathbf{X}^{(2)}\|$ defined from the underlying Gaussian vectors $\mathbf{X}^{(1)} \in N_n(0,\sigma_1^2)$ and $\mathbf{X}^{(2)} \in N_n(0,\sigma_2^2)$. Recalling that only components of these vectors with identical subscripts are correlated with covariance matrix as in (1.10), then the joint PDF and CDF of R_1 and R_2 are given by

$$p_{R_1,R_2}(r_1,r_2) = \frac{(r_1 r_2)^{n/2}}{(1-\rho^2)(\sigma_1\sigma_2)^{n/2+1}(2|\rho|)^{n/2-1}\Gamma(n/2)} \exp\left[-\frac{1}{2(1-\rho^2)}\left(\frac{r_1^2}{\sigma_1^2}+\frac{r_2^2}{\sigma_2^2}\right)\right]$$

$$\times I_{n/2-1}\left(\frac{r_1 r_2 |\rho|}{\sigma_1\sigma_2(1-\rho^2)}\right), r_1 \geq 0, r_2 \geq 0$$

$$(3.6)$$

$$P_{R_1,R_2}(r_1,r_2) = \frac{(1-\rho^2)^{n/2}}{\Gamma(n/2)} \sum_{k=0}^{\infty} \rho^{2k} \frac{\gamma\left(\frac{n}{2}+k,\frac{r_1^2}{2\sigma_1^2(1-\rho^2)}\right)\gamma\left(\frac{n}{2}+k,\frac{r_2^2}{2\sigma_2^2(1-\rho^2)}\right)}{k!\Gamma\left(\frac{n}{2}+k\right)},$$

$$r_1 \geq 0, r_2 \geq 0$$

$$(3.7)$$

The joint moments of R_1 and R_2 are given by

$$E\{R_1^{k_1} R_2^{k_2}\} = \frac{2^{(k_1+k_2)/2}\Gamma\left(\frac{n+k_1}{2}\right)\Gamma\left(\frac{n+k_2}{2}\right)\sigma_1^{k_1}\sigma_2^{k_2}(1-\rho^2)^{n+(k_1+k_2)/2}}{\Gamma^2(n/2)}$$

$$\times {}_2F_1\left(\frac{n+k_1}{2},\frac{n+k_2}{2};\frac{n}{2};\rho^2\right), k_1,k_2 \text{ integer}$$

$$(3.8)$$

where ${}_2F_1(\alpha,\beta;\gamma;x)$ is the Gaussian hypergeometric function [2].

1. $n = 2$

$$p_{R_1,R_2}(r_1,r_2) = \frac{r_1 r_2}{\sigma_1^2 \sigma_2^2 (1-\rho^2)} \exp\left[-\frac{1}{2(1-\rho^2)}\left(\frac{r_1^2}{\sigma_1^2} + \frac{r_2^2}{\sigma_2^2}\right)\right] I_0\left(\frac{r_1 r_2 |\rho|}{\sigma_1 \sigma_2 (1-\rho^2)}\right),$$

$$r_1 \geq 0, r_2 \geq 0$$

(3.9)

$$P_{R_1,R_2}(r_1,r_2) = 1 - \exp\left(-\frac{r_1^2}{2\sigma_1^2}\right) Q_1\left(\sqrt{\frac{r_2^2}{\sigma_2^2(1-\rho^2)}}, \sqrt{\frac{\rho^2 r_1^2}{\sigma_1^2(1-\rho^2)}}\right)$$

$$- \exp\left(-\frac{r_2^2}{2\sigma_2^2}\right)\left[1 - Q_1\left(\sqrt{\frac{\rho^2 r_2^2}{\sigma_2^2(1-\rho^2)}}, \sqrt{\frac{r_1^2}{\sigma_1^2(1-\rho^2)}}\right)\right], \quad (3.10)$$

$$r_1 \geq 0, r_2 \geq 0$$

Using the alternative representation of the first-order Marcum Q-function given in Eqs. (A.6) and (A.7) of Appendix A, the bivariate Rayleigh CDF of (3.10) can be expressed in the form of a single integral with finite limits as follows :

$$P_{R_1,R_2}(r_1,r_2) = 1 - g(Y_1,Y_2|\rho) + \frac{1}{2\pi}\int_{-\pi}^{\pi} \exp\left\{-\frac{Y_1^2 + Y_2^2 + 2|\rho|Y_1 Y_2 \sin\theta}{1-\rho^2}\right\}$$

$$\times \left[\frac{(1-\rho^4)Y_1^2 Y_2^2 + |\rho|(1-\rho^2)Y_1 Y_2(Y_1^2 + Y_2^2)\sin\theta}{(\rho^2 Y_1^2 + 2|\rho|Y_1 Y_2 \sin\theta + Y_2^2)(Y_1^2 + 2|\rho|Y_1 Y_2 \sin\theta + \rho^2 Y_2^2)}\right] d\theta, \quad (3.11)$$

$$r_1 \geq 0, r_2 \geq 0$$

where $Y_i \triangleq r_i / \sqrt{2}\sigma_i$ and

$$g(Y_1,Y_2|\rho) = \begin{cases} \exp\{-Y_2^2\}, & 0 \leq Y_2 < |\rho|Y_1 \\ \frac{1}{2}\exp\{-Y_1^2\} + \exp\{-\rho^2 Y_1^2\}, & Y_2 = |\rho|Y_1 \\ \exp\{-Y_1^2\} + \exp\{-Y_2^2\}, & |\rho|Y_1 < Y_2 < Y_1/|\rho| \\ \frac{1}{2}\exp\{-Y_2^2\} + \exp\{-\rho^2 Y_2^2\}, & Y_2 = Y_1/|\rho| \\ \exp\{-Y_1^2\}, & Y_1/|\rho| < Y_2 \end{cases} \quad (3.12)$$

C. Rician

Consider the pair of Rician RVs of order n, $R_1 = \left\| \mathbf{X}^{(1)} \right\|$, $R_2 = \left\| \mathbf{X}^{(2)} \right\|$ defined from the underlying Gaussian vectors $\mathbf{X}^{(1)} \in N_n\left(\overline{\mathbf{X}}^{(1)}, \sigma_1^2\right)$ and $\mathbf{X}^{(2)} \in N_n\left(\overline{\mathbf{X}}^{(2)}, \sigma_2^2\right)$ where in addition $\left\| \overline{\mathbf{X}}^{(1)} \right\| = \left\| \overline{\mathbf{X}}^{(2)} \right\| = a$. Again recalling that only components of these vectors with identical subscripts are correlated with covariance matrix as in (1.10), then the joint PDF of R_1 and R_2 is given (for $n > 2$) by

$$
p_{R_1,R_2}(r_1,r_2) = \frac{r_1 r_2 \Gamma\left(\dfrac{n}{2}-1\right)}{\left[\sigma_1^2 \sigma_2^2\left(1-\rho^2\right)\right]^{n/2}} \left(-\frac{2\left(1-\rho^2\right)^3 \sigma_1^3 \sigma_2^3}{a^2 \rho\left(1-\rho\sigma_2/\sigma_1\right)\left(1-\rho\sigma_1/\sigma_2\right)}\right)^{n/2-1}
$$

$$
\times \exp\left[-\frac{1}{2\left(1-\rho^2\right)}\left(\frac{r_1^2}{\sigma_1^2}+\frac{r_2^2}{\sigma_2^2}+\left(\frac{\sigma_1^2+\sigma_2^2-2\rho\sigma_1\sigma_2}{\sigma_1^2\sigma_2^2}\right)a^2\right)\right]
$$

$$
\times \sum_{i=0}^{\infty}(-1)^i\left(\frac{n}{2}+i-1\right)\binom{n+i-3}{n-3} I_{n/2+i-1}\left(-\frac{r_1 r_2 \rho}{\left(1-\rho^2\right)\sigma_1\sigma_2}\right)
$$

$$
\times I_{n/2+i-1}\left(\frac{a r_1\left(1-\rho\sigma_1/\sigma_2\right)}{\left(1-\rho^2\right)\sigma_1^2}\right) I_{n/2+i-1}\left(\frac{a r_2\left(1-\rho\sigma_2/\sigma_1\right)}{\left(1-\rho^2\right)\sigma_2^2}\right), r_1 \geq 0, r_2 \geq 0
$$

$$(3.13)$$

For the special case of $n = 2$, the joint PDF is given by

$$
p_{R_1,R_2}(r_1,r_2) = \frac{r_1 r_2}{\sigma_1^2 \sigma_2^2\left(1-\rho^2\right)}
$$

$$
\times \exp\left[-\frac{1}{2\left(1-\rho^2\right)}\left(\frac{r_1^2}{\sigma_1^2}+\frac{r_2^2}{\sigma_2^2}+\left(\frac{\sigma_1^2+\sigma_2^2-2\rho\sigma_1\sigma_2}{\sigma_1^2\sigma_2^2}\right)a^2\right)\right]
$$

$$
\times \sum_{i=0}^{\infty}\varepsilon_i I_i\left(\frac{r_1 r_2 \rho}{\left(1-\rho^2\right)\sigma_1\sigma_2}\right) I_i\left(\frac{a r_1\left(1-\rho\sigma_1/\sigma_2\right)}{\left(1-\rho^2\right)\sigma_1^2}\right)
$$

$$
\times I_i\left(\frac{a r_2\left(1-\rho\sigma_2/\sigma_1\right)}{\left(1-\rho^2\right)\sigma_2^2}\right), r_1 \geq 0, r_2 \geq 0
$$

where ε_i is the Neumann factor, i.e., $\varepsilon_0 = 1$, $\varepsilon_i = 2$ for $i > 0$.

D. Central Chi-Square

Consider the pair of central chi-square RVs of order n, $Y_1 = \|\mathbf{X}^{(1)}\|^2$, $Y_2 = \|\mathbf{X}^{(2)}\|^2$ defined from the underlying Gaussian vectors $\mathbf{X}^{(1)} \in N_n(0, \sigma_1^2)$ and $\mathbf{X}^{(2)} \in N_n(0, \sigma_2^2)$. Then, the joint PDF and CDF of Y_1 and Y_2 are given by

$$p_{Y_1,Y_2}(y_1, y_2) = \frac{(y_1 y_2)^{(n/2-1)/2}}{4(1-\rho^2)(\sigma_1 \sigma_2)^{n/2+1}(2|\rho|)^{n/2-1}\Gamma(n/2)} \exp\left[-\frac{1}{2(1-\rho^2)}\left(\frac{y_1}{\sigma_1^2} + \frac{y_2}{\sigma_2^2}\right)\right]$$

$$\times I_{n/2-1}\left(\frac{\sqrt{y_1 y_2}|\rho|}{\sigma_1 \sigma_2(1-\rho^2)}\right), \quad y_1 \geq 0, y_2 \geq 0$$

$$(3.14)$$

$$P_{Y_1,Y_2}(y_1, y_2) = \frac{(1-\rho^2)^{n/2}}{\Gamma(n/2)} \sum_{k=0}^{\infty} \rho^{2k} \frac{\gamma\left(\frac{n}{2}+k, \frac{y_1}{2\sigma_1^2(1-\rho^2)}\right)\gamma\left(\frac{n}{2}+k, \frac{y_2}{2\sigma_2^2(1-\rho^2)}\right)}{k!\Gamma\left(\frac{n}{2}+k\right)},$$

$$y_1 \geq 0, y_2 \geq 0$$

$$(3.15)$$

The joint moments of Y_1 and Y_2 are given by

$$E\{Y_1^{k_1} Y_2^{k_2}\} = \frac{2^{(k_1+k_2)}\Gamma\left(\frac{n}{2}+k_1\right)\Gamma\left(\frac{n}{2}+k_2\right)\sigma_1^{2k_1}\sigma_2^{2k_2}(1-\rho^2)^{n+k_1+k_2}}{\Gamma^2(n/2)}$$

$$(3.16)$$

$$\times {}_2F_1\left(\frac{n}{2}+k_1, \frac{n}{2}+k_2; \frac{n}{2}; \rho^2\right), \quad k_1, k_2 \text{ integer}$$

1. $n = 2$

$$p_{Y_1,Y_2}(y_1, y_2) = \frac{1}{4\sigma_1^2\sigma_2^2(1-\rho^2)} \exp\left[-\frac{1}{2(1-\rho^2)}\left(\frac{y_1}{\sigma_1^2} + \frac{y_2}{\sigma_2^2}\right)\right] I_0\left(\frac{\sqrt{y_1 y_2}|\rho|}{\sigma_1 \sigma_2(1-\rho^2)}\right),$$

$$y_1 \geq 0, y_2 \geq 0$$

$$(3.17)$$

$$P_{Y_1,Y_2}(y_1,y_2) = 1 - \exp\left(-\frac{y_1}{2\sigma_1^2}\right) Q_1\left(\sqrt{\frac{y_2}{\sigma_2^2(1-\rho^2)}}, \sqrt{\frac{\rho^2 y_1}{\sigma_1^2(1-\rho^2)}}\right)$$

$$-\exp\left(-\frac{y_2}{2\sigma_2^2}\right)\left[1 - Q_1\left(\sqrt{\frac{\rho^2 y_2}{\sigma_2^2(1-\rho^2)}}, \sqrt{\frac{y_1}{\sigma_1^2(1-\rho^2)}}\right)\right], \ y_1 \geq 0, y_2 \geq 0$$

$$(3.18)$$

E. Noncentral Chi-Square

Consider the pair of central chi-square RVs of order n, $Y_1 = \|\mathbf{X}^{(1)}\|^2, Y_2 = \|\mathbf{X}^{(2)}\|^2$ defined from the underlying Gaussian vectors $\mathbf{X}^{(1)} \in N_n(\overline{\mathbf{X}}^{(1)}, \sigma_1^2)$ and $\mathbf{X}^{(2)} \in N_n(\overline{\mathbf{X}}^{(2)}, \sigma_2^2)$. Then, the joint PDF of Y_1 and Y_2 is given (for $n > 2$) by

$$p_{Y_1,Y_2}(y_1,y_2) = \frac{\Gamma\left(\frac{n}{2}-1\right)}{4\left[\sigma_1^2\sigma_2^2(1-\rho^2)\right]^{n/2}}\left(-\frac{2(1-\rho^2)^3\sigma_1^3\sigma_2^3}{a^2\rho(1-\rho\sigma_2/\sigma_1)(1-\rho\sigma_1/\sigma_2)}\right)^{n/2-1}$$

$$\times \exp\left[-\frac{1}{2(1-\rho^2)}\left(\frac{y_1}{\sigma_1^2} + \frac{y_2}{\sigma_2^2} + \left(\frac{\sigma_1^2 + \sigma_2^2 - 2\rho\sigma_1\sigma_2}{\sigma_1^2\sigma_2^2}\right)a^2\right)\right]$$

$$\times \sum_{i=0}^{\infty}(-1)^i\left(\frac{n}{2}+i-1\right)\binom{n+i-3}{n-3}I_{n/2+i-1}\left(-\frac{\sqrt{y_1 y_2}\rho}{(1-\rho^2)\sigma_1\sigma_2}\right) \qquad (3.19)$$

$$\times I_{n/2+i-1}\left(\frac{a\sqrt{y_1}(1-\rho\sigma_1/\sigma_2)}{(1-\rho^2)\sigma_1^2}\right)I_{n/2+i-1}\left(\frac{a\sqrt{y_2}(1-\rho\sigma_2/\sigma_1)}{(1-\rho^2)\sigma_2^2}\right),$$

$$y_1 \geq 0, y_2 \geq 0$$

For the special case of $n = 2$, the joint PDF is given by

$$p_{Y_1,Y_2}(y_1,y_2) = \frac{1}{4\sigma_1^2\sigma_2^2(1-\rho^2)}$$

$$\times \exp\left[-\frac{1}{2(1-\rho^2)}\left(\frac{y_1}{\sigma_1^2} + \frac{y_2}{\sigma_2^2} + \left(\frac{\sigma_1^2 + \sigma_2^2 - 2\rho\sigma_1\sigma_2}{\sigma_1^2\sigma_2^2}\right)a^2\right)\right]$$

$$\times \sum_{i=0}^{\infty} \varepsilon_i I_i \left(\frac{\sqrt{y_1 y_2} \rho}{(1-\rho^2)\sigma_1\sigma_2} \right) I_i \left(\frac{a\sqrt{y_1}(1-\rho\sigma_1/\sigma_2)}{(1-\rho^2)\sigma_1^2} \right)$$

$$\times I_i \left(\frac{a\sqrt{y_2}(1-\rho\sigma_2/\sigma_1)}{(1-\rho^2)\sigma_2^2} \right), y_1 \geq 0, y_2 \geq 0$$

F. Log-Normal

Consider the pair of correlated log-normal RVs $\gamma_1 = 10^{X_1/10}$ and $\gamma_2 = 10^{X_2/10}$ corresponding to the correlated Gaussian RVs $X_1 \in N_1(\overline{X}_1, \sigma_1^2)$ and $X_2 \in N_1(\overline{X}_2, \sigma_2^2)$. Then the joint PDF of γ_1 and γ_2 is

$$p_{\gamma_1,\gamma_2}(\gamma_1,\gamma_2) = \frac{\xi^2}{2\pi\sigma_1\sigma_2\sqrt{1-\rho^2}\gamma_1\gamma_2} \exp\left\{ -\frac{1}{2(1-\rho^2)} \left[\left(\frac{10\log_{10}\gamma_1 - \overline{X}_1}{\sigma_1} \right)^2 \right. \right.$$

$$\left. \left. + \left(\frac{10\log_{10}\gamma_2 - \overline{X}_2}{\sigma_2} \right)^2 - 2\rho \left(\frac{10\log_{10}\gamma_1 - \overline{X}_1}{\sigma_1} \right) \left(\frac{10\log_{10}\gamma_2 - \overline{X}_2}{\sigma_2} \right) \right] \right\}$$

$$(3.20)$$

Once again the joint CDF and also the joint CF are not available in closed form. However, the joint moments can be determined from the joint CDF of the corresponding Gaussian RVs. In particular,

$$\Psi_{X_1,X_2}(\omega_1,\omega_2) = E\{e^{j\omega_1 X_1 + j\omega_2 X_2}\} = E\{e^{j\omega_1\xi\ln\gamma_1 + j\omega_2\xi\ln\gamma_2}\} = E\{\gamma_1^{j\omega_1\xi}\gamma_2^{j\omega_2\xi}\} \quad (3.21)$$

Since from (3.4)

$$\Psi_{X_1,X_2}(\omega_1,\omega_2) = \exp\left(j\omega_1\overline{X}_1 + j\omega_2\overline{X}_2 - \frac{1}{2}\omega_1^2\sigma_1^2 - \frac{1}{2}\omega_2^2\sigma_2^2 - \rho\omega_1\omega_2\sigma_1\sigma_2 \right)$$

$$(3.22)$$

then

$$E\{\gamma_1^{k_1}\gamma_2^{k_2}\} = \exp\left(\frac{k_1\overline{X}_1}{\xi} + \frac{k_2\overline{X}_2}{\xi} + \frac{k_1^2\sigma_1^2}{2\xi^2} + \frac{k_2^2\sigma_2^2}{2\xi^2} + \frac{\rho k_1 k_2 \sigma_1 \sigma_2}{\xi^2} \right)$$

$$= E\{\gamma_1^{k_1}\}E\{\gamma_2^{k_2}\}\exp\left(\frac{\rho k_1 k_2 \sigma_1 \sigma_2}{\xi^2} \right)$$

$$(3.23)$$

4

DIFFERENCE OF CHI-SQUARE RANDOM VARIABLES

A. Independent Central Chi-Square (-) Central Chi-Square

Define $Y = Y_1 - Y_2$ where Y_1 and Y_2 are independent central chi-square distributed RVs with n_1 and n_2 degrees of freedom, respectively.

1. $n_1 = n_2 = 1$

$$P_Y(y) = \frac{1}{2\pi\sigma_1\sigma_2}\exp\left(-\frac{\sigma_2^2-\sigma_1^2}{4\sigma_1^2\sigma_2^2}y\right)K_0\left(\frac{\sigma_2^2+\sigma_1^2}{4\sigma_1^2\sigma_2^2}|y|\right) \qquad (4.1)$$

$$P_Y(y) = ------ \qquad (4.2)$$

$$\Psi_Y(\omega) = \left(\frac{1}{(1-2j\omega\sigma_1^2)(1+2j\omega\sigma_2^2)}\right)^{1/2} \qquad (4.3)$$

where $K_0(x)$ is the modified Bessel function of the second kind [2].[3]

2. $n_1 = n_2 = 2$

---

[3] Note that $K_0(x)$ is defined for $0 < x \le \infty$ whereas $I_0(x)$ is defined for $-\infty \le x \le \infty$ and is an even function of x.

$$p_Y(y) = \begin{cases} \dfrac{1}{2(\sigma_1^2 + \sigma_2^2)} \exp\left(\dfrac{y}{2\sigma_2^2}\right), & y < 0 \\[4mm] \dfrac{1}{2(\sigma_1^2 + \sigma_2^2)} \exp\left(-\dfrac{y}{2\sigma_1^2}\right), & y \geq 0 \end{cases} \qquad (4.4)$$

$$P_Y(y) = \begin{cases} \dfrac{\sigma_2^2}{\sigma_1^2 + \sigma_2^2} \exp\left(\dfrac{y}{2\sigma_2^2}\right), & y < 0 \\[4mm] 1 - \dfrac{\sigma_1^2}{\sigma_1^2 + \sigma_2^2} \exp\left(-\dfrac{y}{2\sigma_1^2}\right), & y \geq 0 \end{cases} \qquad (4.5)$$

$$\Psi_Y(\omega) = \frac{1}{(1 - 2j\omega\sigma_1^2)(1 + 2j\omega\sigma_2^2)} \qquad (4.6)$$

3. $n_1 = n_2 = 2m$

$$p_Y(y) = \begin{cases} \dfrac{1}{2\sigma_2^2} \exp\left(\dfrac{y}{2\sigma_2^2}\right) \dfrac{1}{(m-1)!} \left(\dfrac{\sigma_2^2}{\sigma_1^2 + \sigma_2^2}\right)^m \sum\limits_{i=0}^{m-1} \dfrac{(2(m-1)-i)!}{i!(m-1-i)!} \\[3mm] \times \left(\dfrac{\sigma_1^2}{\sigma_1^2 + \sigma_2^2}\right)^{m-1-i} \left(-\dfrac{y}{2\sigma_2^2}\right)^i, \quad y < 0 \\[5mm] \dfrac{1}{2\sigma_1^2} \exp\left(-\dfrac{y}{2\sigma_1^2}\right) \dfrac{1}{(m-1)!} \left(\dfrac{\sigma_1^2}{\sigma_1^2 + \sigma_2^2}\right)^m \sum\limits_{i=0}^{m-1} \dfrac{(2(m-1)-i)!}{i!(m-1-i)!} \\[3mm] \times \left(\dfrac{\sigma_2^2}{\sigma_1^2 + \sigma_2^2}\right)^{m-1-i} \left(\dfrac{y}{2\sigma_1^2}\right)^i, \quad y \geq 0 \end{cases}$$

$$(4.7)$$

$$P_Y(y) = \exp\left(\frac{y}{2\sigma_2^2}\right) \frac{1}{(m-1)!} \left(\frac{\sigma_2^2}{\sigma_1^2 + \sigma_2^2}\right)^m \sum_{i=0}^{m-1} \sum_{l=0}^{i} \frac{(2(m-1)-i)!}{(i-l)!(m-1-i)!}$$
$$\times \left(\frac{\sigma_1^2}{\sigma_1^2 + \sigma_2^2}\right)^{m-1-i} \left(-\frac{y}{2\sigma_2^2}\right)^{i-l}, \quad y < 0$$

$$P_Y(y) = 1 - \exp\left(-\frac{y}{2\sigma_1^2}\right)\frac{1}{(m-1)!}\left(\frac{\sigma_1^2}{\sigma_1^2+\sigma_2^2}\right)^m \sum_{i=0}^{m-1}\sum_{l=0}^{i}\frac{(2(m-1)-i)!}{(i-l)!(m-1-i)!}$$

$$\times \left(\frac{\sigma_2^2}{\sigma_1^2+\sigma_2^2}\right)^{m-1-i}\left(\frac{y}{2\sigma_1^2}\right)^{i-l}, \; y \geq 0 \tag{4.8}$$

$$\Psi_Y(\omega) = \left(\frac{1}{(1-2j\omega\sigma_1^2)(1+2j\omega\sigma_2^2)}\right)^m \tag{4.9}$$

4. $n_1 = n_2 = 2m+1$

$$p_Y(y) = \frac{1}{2\sqrt{\pi}\sigma_1\sigma_2\Gamma(m+1/2)}\left(\frac{|y|}{2(\sigma_2^2+\sigma_1^2)}\right)^m \exp\left(-\frac{\sigma_2^2-\sigma_1^2}{4\sigma_1^2\sigma_2^2}y\right)$$

$$\times K_m\left(\frac{\sigma_2^2+\sigma_1^2}{4\sigma_1^2\sigma_2^2}|y|\right) \tag{4.10}$$

$$P_Y(y) = ------ \tag{4.11}$$

$$\Psi_Y(\omega) = \left(\frac{1}{(1-2j\omega\sigma_1^2)(1+2j\omega\sigma_2^2)}\right)^{m+1/2} \tag{4.12}$$

5. $n_1 = 2m, n_2 = 2$

$$p_Y(y) = \begin{cases} \dfrac{1}{2\sigma_2^2}\left(\dfrac{\sigma_2^2}{\sigma_1^2+\sigma_2^2}\right)^m \exp\left(\dfrac{y}{2\sigma_2^2}\right), & y < 0 \\[3mm] \dfrac{1}{2\sigma_1^2}\exp\left(-\dfrac{y}{2\sigma_1^2}\right)\left(\dfrac{\sigma_1^2}{\sigma_1^2+\sigma_2^2}\right)\sum_{i=0}^{m-1}\dfrac{1}{i!}\left(\dfrac{\sigma_2^2}{\sigma_1^2+\sigma_2^2}\right)^{m-1-i}\left(\dfrac{y}{2\sigma_1^2}\right)^i, & y \geq 0 \end{cases} \tag{4.13}$$

$$P_Y(y) = \begin{cases} \left(\dfrac{\sigma_2^2}{\sigma_1^2+\sigma_2^2}\right)^m \exp\left(\dfrac{y}{2\sigma_2^2}\right), & y < 0 \\[3mm] 1-\exp\left(-\dfrac{y}{2\sigma_1^2}\right)\left(\dfrac{\sigma_1^2}{\sigma_1^2+\sigma_2^2}\right)\sum_{i=0}^{m-1}\sum_{l=0}^{i}\dfrac{1}{(i-l)!}\left(\dfrac{\sigma_2^2}{\sigma_1^2+\sigma_2^2}\right)^{m-1-i}\left(\dfrac{y}{2\sigma_1^2}\right)^{i-l}, & y \geq 0 \end{cases} \tag{4.14}$$

$$\Psi_Y(\omega) = \frac{1}{\left(1 - 2j\omega\sigma_1^2\right)^m \left(1 + 2j\omega\sigma_2^2\right)} \tag{4.15}$$

6. $n_1 = 2m_1, n_2 = 2m_2$

$$p_Y(y) = \begin{cases} \dfrac{1}{2\sigma_2^2}\exp\left(\dfrac{y}{2\sigma_2^2}\right)\dfrac{1}{(m_1-1)!}\left(\dfrac{\sigma_2^2}{\sigma_1^2+\sigma_2^2}\right)^{m_1}\displaystyle\sum_{i=0}^{m_2-1}\dfrac{(m_1+m_2-2-i)!}{i!(m_2-1-i)!} \\ \quad \times\left(\dfrac{\sigma_1^2}{\sigma_1^2+\sigma_2^2}\right)^{m_2-1-i}\left(-\dfrac{y}{2\sigma_2^2}\right)^i, \ y<0 \\[3mm] \dfrac{1}{2\sigma_1^2}\exp\left(-\dfrac{y}{2\sigma_1^2}\right)\dfrac{1}{(m_2-1)!}\left(\dfrac{\sigma_1^2}{\sigma_1^2+\sigma_2^2}\right)^{m_2}\displaystyle\sum_{i=0}^{m_1-1}\dfrac{(m_1+m_2-2-i)!}{i!(m_1-1-i)!} \\ \quad \times\left(\dfrac{\sigma_2^2}{\sigma_1^2+\sigma_2^2}\right)^{m_1-1-i}\left(\dfrac{y}{2\sigma_1^2}\right)^i, \ y\geq0 \end{cases} \tag{4.16}$$

$$P_Y(y) = \begin{cases} \exp\left(\dfrac{y}{2\sigma_2^2}\right)\dfrac{1}{(m_1-1)!}\left(\dfrac{\sigma_2^2}{\sigma_1^2+\sigma_2^2}\right)^{m_1}\displaystyle\sum_{i=0}^{m_2-1}\sum_{l=0}^{i}\dfrac{(m_1+m_2-2-i)!}{(i-l)!(m_2-1-i)!} \\ \quad \times\left(\dfrac{\sigma_1^2}{\sigma_1^2+\sigma_2^2}\right)^{m_2-1-i}\left(-\dfrac{y}{2\sigma_2^2}\right)^{i-l}, \ y<0 \\[3mm] 1-\exp\left(-\dfrac{y}{2\sigma_1^2}\right)\dfrac{1}{(m_2-1)!}\left(\dfrac{\sigma_1^2}{\sigma_1^2+\sigma_2^2}\right)^{m_2}\displaystyle\sum_{i=0}^{m_1-1}\sum_{l=0}^{i}\dfrac{(m_1+m_2-2-i)!}{(i-l)!(m_1-1-i)!} \\ \quad \times\left(\dfrac{\sigma_2^2}{\sigma_1^2+\sigma_2^2}\right)^{m_1-1-i}\left(\dfrac{y}{2\sigma_1^2}\right)^{i-l}, \ y\geq0 \end{cases} \tag{4.17}$$

$$\Psi_Y(\omega) = \frac{1}{\left(1 - 2j\omega\sigma_1^2\right)^{m_1}\left(1 + 2j\omega\sigma_2^2\right)^{m_2}} \tag{4.18}$$

B. Dependent Central Chi-Square (-) Central Chi-Square

1. $n_1 = n_2 = 1$

To simplify the expressions, we introduce the parameters

$$\gamma^- = \frac{\left[\left(\sigma_2^2 - \sigma_1^2\right)^2 + 4\sigma_1^2\sigma_2^2\left(1-\rho^2\right)\right]^{1/2}}{\sigma_1^2\sigma_2^2\left(1-\rho^2\right)}, \quad \alpha^\pm = \gamma^- \pm \frac{\sigma_2^2 - \sigma_1^2}{\sigma_1^2\sigma_2^2\left(1-\rho^2\right)} \quad (4.19)$$

Note that $\alpha^+ \geq 0$ and $\alpha^- \geq 0$. Then,

$$p_Y(y) = \frac{1}{2\pi\sigma_1\sigma_2\sqrt{1-\rho^2}} \exp\left(-\frac{1}{4}(\alpha^+ - \gamma^-)y\right) K_0\left(\frac{1}{4}\gamma^-|y|\right) \quad (4.20)$$

$$P_Y(y) = ------ \quad (4.21)$$

$$\Psi_Y(\omega) = \left(\frac{1-\rho^2}{\left(1-2j\omega(1-\rho^2)\sigma_1^2\right)\left(1+2j\omega(1-\rho^2)\sigma_2^2\right) - \rho^2}\right)^{1/2} \quad (4.22)$$

2. $n_1 = n_2 = 2$

$$p_Y(y) = \begin{cases} \dfrac{1}{2\sigma_1^2\sigma_2^2\left(1-\rho^2\right)\gamma^-} \exp\left(\dfrac{1}{4}\alpha^- y\right), & y < 0 \\[4mm] \dfrac{1}{2\sigma_1^2\sigma_2^2\left(1-\rho^2\right)\gamma^-} \exp\left(-\dfrac{1}{4}\alpha^+ y\right), & y \geq 0 \end{cases} \quad (4.23)$$

$$P_Y(y) = \begin{cases} \dfrac{2}{\sigma_1^2\sigma_2^2\left(1-\rho^2\right)\gamma^-\alpha^-} \exp\left(\dfrac{1}{4}\alpha^- y\right), & y < 0 \\[4mm] 1 - \dfrac{2}{\sigma_1^2\sigma_2^2\left(1-\rho^2\right)\gamma^-\alpha^+} \exp\left(-\dfrac{1}{4}\alpha^+ y\right), & y \geq 0 \end{cases} \quad (4.24)$$

$$\Psi_Y(\omega) = \frac{1-\rho^2}{\left(1-2j\omega(1-\rho^2)\sigma_1^2\right)\left(1+2j\omega(1-\rho^2)\sigma_2^2\right) - \rho^2} \quad (4.25)$$

3. $n_1 = n_2 = 2m$

$$p_Y(y) = \frac{|y|^{m-1}}{(m-1)!\left[2\sigma_1^2\sigma_2^2\left(1-\rho^2\right)\gamma^-\right]^m} \exp\left(\frac{1}{4}\alpha^- y\right) \sum_{i=0}^{m-1} \frac{(m+i-1)!}{i!(m-i-1)!}\left(\frac{2}{\gamma^-|y|}\right)^i, \quad y < 0$$

$$p_Y(y) = \frac{|y|^{m-1}}{(m-1)!\left[2\sigma_1^2\sigma_2^2(1-\rho^2)\gamma^-\right]^m} \exp\left(-\frac{1}{4}\alpha^+ y\right) \sum_{i=0}^{m-1} \frac{(m+i-1)!}{i!(m-i-1)!} \left(\frac{2}{\gamma^-|y|}\right)^i,$$

$$y \geq 0$$

$$(4.26)$$

$$P_Y(y) = \begin{cases} \dfrac{1}{(m-1)!\left(2\sigma_1^2\sigma_2^2\gamma^-\right)^m} \exp\left(\dfrac{1}{4}\alpha^- y\right) \\[2mm] \times \displaystyle\sum_{i=0}^{m-1}\sum_{l=0}^{m-i-1} \dfrac{(m+i-1)!}{i!(m-i-l-1)!} \left(\dfrac{2}{\gamma^-}\right)^i \left(\dfrac{4}{\alpha^-}\right)^{l+1} (-y)^{m-i-l-1}, y<0 \\[4mm] 1 - \dfrac{1}{(m-1)!\left(2\sigma_1^2\sigma_2^2\gamma^-\right)^m} \exp\left(-\dfrac{1}{4}\alpha^+ y\right) \\[2mm] \times \displaystyle\sum_{i=0}^{m-1}\sum_{l=0}^{m-i-1} \dfrac{(m+i-1)!}{i!(m-i-l-1)!} \left(\dfrac{2}{\gamma^-}\right)^i \left(\dfrac{4}{\alpha^+}\right)^{l+1} y^{m-i-l-1}, y\geq 0 \end{cases}$$

$$(4.27)$$

$$\Psi_Y(\omega) = \left(\frac{1-\rho^2}{\left(1-2j\omega(1-\rho^2)\sigma_1^2\right)\left(1+2j\omega(1-\rho^2)\sigma_2^2\right)-\rho^2}\right)^m \qquad (4.28)$$

4. $n_1 = n_2 = 2m+1$

$$p_Y(y) = \frac{\sqrt{\gamma^-/2}\,|y|^m}{\sqrt{\pi}\,\Gamma(m+1/2)\left[2\sigma_1^2\sigma_2^2(1-\rho^2)\gamma^-\right]^{m+1/2}} \exp\left(-\frac{1}{4}(\alpha^+-\gamma^-)y\right) K_m\left(\frac{\gamma^-|y|}{4}\right)$$

$$(4.29)$$

$$P_Y(y) = - - - - - - \qquad (4.30)$$

$$\Psi_Y(\omega) = \left(\frac{1-\rho^2}{\left(1-2j\omega(1-\rho^2)\sigma_1^2\right)\left(1+2j\omega(1-\rho^2)\sigma_2^2\right)-\rho^2}\right)^{m+1/2} \qquad (4.31)$$

C. Independent Noncentral Chi-Square (-) Central Chi-Square

Define $Y = Y_1 - Y_2$ where Y_1 and Y_2 are independent noncentral and central chi-square distributed RVs with n_1 and n_2 degrees of freedom, respectively.

1. $n_1 = n_2 = 2$

$$
p_Y(y) = \begin{cases}
\dfrac{1}{2(\sigma_1^2 + \sigma_2^2)} \exp\left(\dfrac{y}{2\sigma_2^2}\right) \exp\left(-\dfrac{a_1^2}{2(\sigma_1^2 + \sigma_2^2)}\right), & y < 0 \\[3mm]
\dfrac{1}{2(\sigma_1^2 + \sigma_2^2)} \exp\left(\dfrac{y}{2\sigma_2^2}\right) \exp\left(-\dfrac{a_1^2}{2(\sigma_1^2 + \sigma_2^2)}\right) \\[3mm]
\quad \times Q_1\left(\dfrac{a_1}{\sigma_1}\sqrt{\dfrac{\sigma_2^2}{\sigma_1^2 + \sigma_2^2}}, \sqrt{\dfrac{y(\sigma_1^2 + \sigma_2^2)}{\sigma_1^2 \sigma_2^2}}\right), & y \geq 0
\end{cases}
\tag{4.32}
$$

$$
p_Y(y) = \begin{cases}
\left(\dfrac{\sigma_2^2}{\sigma_1^2 + \sigma_2^2}\right) \exp\left(\dfrac{y}{2\sigma_2^2}\right) \exp\left(-\dfrac{a_1^2}{2(\sigma_1^2 + \sigma_2^2)}\right), & y < 0 \\[3mm]
1 - Q_1\left(\dfrac{a_1}{\sigma_1}, \sqrt{\dfrac{y}{\sigma_1^2}}\right) + \left(\dfrac{\sigma_2^2}{\sigma_1^2 + \sigma_2^2}\right) \exp\left(\dfrac{y}{2\sigma_2^2}\right) \exp\left(-\dfrac{a_1^2}{2(\sigma_1^2 + \sigma_2^2)}\right) \\[3mm]
\quad \times Q_1\left(\dfrac{a_1}{\sigma_1}\sqrt{\dfrac{\sigma_2^2}{\sigma_1^2 + \sigma_2^2}}, \sqrt{\dfrac{y(\sigma_1^2 + \sigma_2^2)}{\sigma_1^2 \sigma_2^2}}\right), & y \geq 0
\end{cases}
\tag{4.33}
$$

$$
\Psi_Y(\omega) = \dfrac{1}{(1 - 2j\omega\sigma_1^2)(1 + 2j\omega\sigma_2^2)} \exp\left(\dfrac{j\omega a^2}{1 - 2j\omega\sigma_1^2}\right)
\tag{4.34}
$$

2. $n_1 = 2m, n_2 = 2$

$$
p_Y(y) = \begin{cases}
\dfrac{1}{2\sigma_2^2}\left(\dfrac{\sigma_2^2}{\sigma_1^2 + \sigma_2^2}\right)^m \exp\left(\dfrac{y}{2\sigma_2^2}\right) \exp\left(-\dfrac{a_1^2}{2(\sigma_1^2 + \sigma_2^2)}\right), & y < 0 \\[3mm]
\dfrac{1}{2\sigma_2^2}\left(\dfrac{\sigma_2^2}{\sigma_1^2 + \sigma_2^2}\right)^m \exp\left(\dfrac{y}{2\sigma_2^2}\right) \exp\left(-\dfrac{a_1^2}{2(\sigma_1^2 + \sigma_2^2)}\right) \\[3mm]
\quad \times Q_m\left(\dfrac{a_1}{\sigma_1}\sqrt{\dfrac{\sigma_2^2}{\sigma_1^2 + \sigma_2^2}}, \sqrt{\dfrac{y(\sigma_1^2 + \sigma_2^2)}{\sigma_1^2 \sigma_2^2}}\right), & y \geq 0
\end{cases}
\tag{4.35}
$$

$$P_Y(y) = \begin{cases} \left(\dfrac{\sigma_2^2}{\sigma_1^2+\sigma_2^2}\right)^m \exp\left(\dfrac{y}{2\sigma_2^2}\right)\exp\left(-\dfrac{a_1^2}{2\left(\sigma_1^2+\sigma_2^2\right)}\right), & y<0 \\[3mm] 1-Q_m\left(\dfrac{a_1}{\sigma_1},\sqrt{\dfrac{y}{\sigma_1^2}}\right)+\left(\dfrac{\sigma_2^2}{\sigma_1^2+\sigma_2^2}\right)^m \exp\left(\dfrac{y}{2\sigma_2^2}\right)\exp\left(-\dfrac{a_1^2}{2\left(\sigma_1^2+\sigma_2^2\right)}\right) \\[3mm] \times Q_m\left(\dfrac{a_1}{\sigma_1}\sqrt{\dfrac{\sigma_2^2}{\sigma_1^2+\sigma_2^2}},\sqrt{\dfrac{y\left(\sigma_1^2+\sigma_2^2\right)}{\sigma_1^2\sigma_2^2}}\right), & y\geq 0 \end{cases}$$ (4.36)

$$\Psi_Y(\omega) = \frac{1}{\left(1-2j\omega\sigma_1^2\right)^m\left(1-2j\omega\sigma_2^2\right)}\exp\left(\frac{j\omega a_1^2}{1-2j\omega\sigma_1^2}\right)$$ (4.37)

3. $n_1 = 2m_1, n_2 = 2m_2$

$$p_Y(y) = \begin{cases} \dfrac{1}{2\sigma_2^2}\left(\dfrac{\sigma_2^2}{\sigma_1^2+\sigma_2^2}\right)^{m_1}\exp\left(\dfrac{y}{2\sigma_2^2}\right)\exp\left(-\dfrac{a_1^2}{2\sigma_1^2}\right)\displaystyle\sum_{i=0}^{m_2-1}\dfrac{1}{i!(m_2-1-i)!}\dfrac{(m_1-1+i)!}{(m_1-1)!} \\[3mm] \times\left(-\dfrac{y}{2\sigma_2^2}\right)^{m_2-1-i}\left(\dfrac{\sigma_1^2}{\sigma_1^2+\sigma_2^2}\right)^i {}_1F_1\left(m_1+i;m_1;\dfrac{a_1^2}{2\sigma_1^2}\left(\dfrac{\sigma_2^2}{\sigma_1^2+\sigma_2^2}\right)\right), & y<0 \\[3mm] \dfrac{1}{2\sigma_2^2}\left(\dfrac{\sigma_2^2}{\sigma_1^2+\sigma_2^2}\right)^{(m_1+1)/2}\exp\left(\dfrac{y}{2\sigma_2^2}\right)\exp\left(-\dfrac{a_1^2}{2\left(\sigma_1^2+\sigma_2^2\right)}\right) \\[3mm] \times\left(\dfrac{\sigma_1^2}{a_1^2}\right)^{(m_1-1)/2}\displaystyle\sum_{i=0}^{m_2-1}\dfrac{1}{i!(m_2-1-i)!2^i}\left(-\dfrac{y}{2\sigma_2^2}\right)^{m_2-1-i}\left(\dfrac{\sigma_1^2}{\sigma_1^2+\sigma_2^2}\right)^i \\[3mm] \times Q_{2i+m_1,m_1-1}\left(\dfrac{a_1}{\sigma_1}\sqrt{\dfrac{\sigma_2^2}{\sigma_1^2+\sigma_2^2}},\sqrt{\dfrac{y\left(\sigma_1^2+\sigma_2^2\right)}{\sigma_1^2\sigma_2^2}}\right), & y\geq 0 \end{cases}$$

(4.38)

$$P_Y(y) = ------$$ (4.39)

$$\Psi_Y(\omega) = \frac{1}{\left(1-2j\omega\sigma_1^2\right)^{m_1}\left(1+2j\omega\sigma_2^2\right)^{m_2}}\exp\left(\frac{j\omega a_1^2}{1-2j\omega\sigma_1^2}\right)$$ (4.40)

where

$$Q_{mn}(\alpha,\beta) = \int_{\beta}^{\infty} x^m \exp\left(-\frac{x^2+\alpha^2}{2}\right) I_n(\alpha x) dx \qquad (4.41)$$

is a generalization of the Marcum Q-function defined in [9, Eq. (86)] and which has recursive properties [9, Eq. (87)]

$$Q_{m,n}(\alpha,\beta) = \alpha Q_{m-1,n+1}(\alpha,\beta) + (m+n-1)Q_{m-2,n}(\alpha,\beta)$$
$$+\beta^{m-1}\exp\left(-\frac{\alpha^2+\beta^2}{2}\right) I_n(\alpha\beta) \qquad (4.42)$$

Note that the special case $Q_{m,m-1}(\alpha,\beta)$ is related to the generalized (mth-order) Marcum Q-function of (2.24) by

$$Q_{m,m-1}(\alpha,\beta) = \alpha^{m-1} Q_m(\alpha,\beta) \qquad (4.43)$$

Before concluding this section we point out that for the case $n_1 = n_2 = m, m$ odd, the PDF can be expressed in the form of an infinite series in Whittaker functions [2] which themselves are expressed in terms of the confluent hypergeometric function $_1F_1(\alpha;\beta;\gamma)$. Because of the absence of the functions in standard mathematical software manipulation packages such as Mathematica®, and the complexity of the resulting expressions, their use is somewhat limited in practical applications and thus the author has decided to omit these results for this case. Nevertheless, the CF is still simple and given by

$$\Psi_Y(\omega) = \frac{1}{(1-2j\omega\sigma_1^2)^{m/2}(1+2j\omega\sigma_2^2)^{m/2}} \exp\left(\frac{j\omega a_1^2}{1-2j\omega\sigma_1^2}\right) \qquad (4.44)$$

D. Independent Central Chi-Square (-) Noncentral Chi-Square

Define $Z = Y_1 - Y_2$ where Y_1 and Y_2 are independent central and noncentral chi-square distributed RVs with n_1 and n_2 degrees of freedom, respectively. Then, the PDF of Z can be determined in terms of the PDF of Y given in the previous section by   ·

$$P_Z(z) = P_Y(-z)\Big|_{\substack{\sigma_2 \to \sigma_1 \\ \sigma_1 \to \sigma_2 \\ a_1^2 \to a_2^2}} \qquad (4.45)$$

Similarly, the CDF of Z can be determined from the CDF of Y as

$$P_Z(z) = 1 - P_Y(-z)\Big|_{\substack{\sigma_2 \to \sigma_1 \\ \sigma_1 \to \sigma_2 \\ a_1^2 \to a_2^2}} \qquad (4.46)$$

E. Independent Noncentral Chi-Square (-) Noncentral Chi-Square

For this case, the results for the PDF can be expressed in the form of a doubly infinite series in Whittaker functions. Once again, for the reasons stated above, the author has decided to omit these results for this case. Nevertheless, the CF for the generic case $n_1 = m_1, n_2 = m_2$ is still simple and given by

$$\Psi_Y(\omega) = \frac{1}{\left(1 - 2j\omega\sigma_1^2\right)^{m_1/2}\left(1 + 2j\omega\sigma_2^2\right)^{m_2/2}} \exp\left(\frac{j\omega a_1^2}{1 - 2j\omega\sigma_1^2}\right) \exp\left(\frac{j\omega a_2^2}{1 - 2j\omega\sigma_2^2}\right)$$

$$(4.47)$$

5

SUM OF CHI-SQUARE RANDOM VARIABLES

Define the RV $Z_2 = -Y_2$. Then the PDF of Z_2 is given by $p_{Z_2}(z) = p_{Y_2}(-z)$, $z \le 0$. From the form of $p_Y(y)$ for central chi-square RVs, we observe that for n odd, the PDF of Z_2 is given by the PDF of Y_2 with y replaced by z and $-\sigma_2^2$ substituted for σ_2^2. For n even, the PDF of Z_2 is given by the negative of the PDF of Y_2 with y replaced by z and $-\sigma_2^2$ substituted for σ_2^2. From the form of $p_Y(y)$ for noncentral chi-square RVs, we observe that in addition to the above substitutions, $-a_2^2$ must be substituted for a_2^2. For example, for Y_2 a central chi-square RV with $2m_2$ degrees of freedom, the PDF of Z_2 is expressible as

$$p_{Z_2}(z) = p_{Y_2}(-z) = \frac{1}{2^{m_2}(\sigma_2^2)^{m_2}\Gamma(m_2)}(-z)^{m_2-1}\exp\left(\frac{z}{2\sigma_2^2}\right)$$

$$= -\frac{1}{2^{m_2}(-\sigma_2^2)^{m_2}\Gamma(m_2)}z^{m_2-1}\exp\left(-\frac{z}{2(-\sigma_2^2)}\right) \tag{5.1}$$

$$= -p_{Y_2}(z)\big|_{\sigma_2^2 \to -\sigma_2^2}, \ z \le 0$$

that is, we use the expression for the PDF of Y_2 (which applies for $y \ge 0$) but substitute z for y, $-\sigma_2^2$ for σ_2^2, and then take its negative and apply it for $z \le 0$. Similarly, for Y_2 a noncentral chi-square RV with $2m_2$ degrees of freedom, the PDF of Z_2 is expressible as

$$p_{Z_2}(z) = \frac{1}{2\sigma_2^2} \left(\frac{-z}{a_2^2} \right)^{(m_2-1)/2} \exp\left(-\frac{-z+a_2^2}{2\sigma_2^2} \right) I_{m_2-1}\left(\sqrt{\frac{a_2^2(-z)}{\sigma_2^4}} \right)$$

$$= -\frac{1}{2(-\sigma_2^2)} \left(\frac{z}{-a_2^2} \right)^{(m_2-1)/2} \exp\left(-\frac{z-a_2^2}{2(-\sigma_2^2)} \right) I_{m_2-1}\left(\sqrt{\frac{-a_2^2 z}{(-\sigma_2^2)^2}} \right) \qquad (5.2)$$

$$= -p_{Y_2}(z)\Big|_{\substack{\sigma_2^2 \to -\sigma_2^2 \\ a_2^2 \to -a_2^2}}, z \le 0$$

A. Independent Central Chi-Square (+) Central Chi-Square

Define now the RV $Z = Y_1 + Y_2 = Y_1 - Z_2$. Also, for the results of Section 4A, define the notation

$$p_Y(y) = \begin{cases} p_Y^-(y), & y < 0 \\ p_Y^+(y), & y \ge 0 \end{cases} \qquad (5.3)$$

Then, it can be shown that the PDF of Z is given by

$$p_Z(z) = \begin{cases} p_Y^-(z)\Big|_{\sigma_2^2 \to -\sigma_2^2} - p_Y^+(z)\Big|_{\sigma_2^2 \to -\sigma_2^2}, & z \ge 0, n_2 \text{ odd} \\ p_Y^+(z)\Big|_{\sigma_2^2 \to -\sigma_2^2} - p_Y^-(z)\Big|_{\sigma_2^2 \to -\sigma_2^2}, & z \ge 0, n_2 \text{ even} \end{cases} \qquad (5.4)$$

Note that since Z only takes on positive (or zero) values, the PDF of Z is defined only for $z \ge 0$. Similarly, define the notation

$$P_Y(y) = \begin{cases} P_Y^-(y), & y < 0 \\ P_Y^+(y), & y \ge 0 \end{cases} \qquad (5.5)$$

Then, it can be shown from (5.4) that

$$P_Z(z) = \begin{cases} P_Y^-(z)\Big|_{\sigma_2^2 \to -\sigma_2^2} - P_Y^+(z)\Big|_{\sigma_2^2 \to -\sigma_2^2}, & z \ge 0, n_2 \text{ odd} \\ P_Y^+(z)\Big|_{\sigma_2^2 \to -\sigma_2^2} - P_Y^-(z)\Big|_{\sigma_2^2 \to -\sigma_2^2}, & z \ge 0, n_2 \text{ even} \end{cases} \qquad (5.6)$$

Before proceeding, the reader is cautioned that care must be exercised in applying (5.4) and (5.6) since in some instances the substitution $\sigma_2^2 \rightarrow -\sigma_2^2$ in the generic form of the PDF of Y might result in functions with imaginary or undefined arguments. In these instances, one is better off deriving the result for the chi-square sum directly from a convolution of the individual chi-square RV PDFs rather than from the result for the chi-square difference. In this same regard, a closed-form result for the CDF might exist for the chi-square sum RV even though it doesn't exist for the chi-square difference RV.

1. $n_1 = n_2 = 1$

$$p_Z(z) = \frac{1}{2\sigma_1\sigma_2} \exp\left(-\frac{\sigma_2^2 + \sigma_1^2}{4\sigma_1^2\sigma_2^2} z \right) I_0\left(\frac{\sigma_2^2 - \sigma_1^2}{4\sigma_1^2\sigma_2^2} z \right), \quad z \geq 0 \qquad (5.7)$$

$$P_Z(z) = 1 + \exp\left(-\frac{\sigma_2^2 + \sigma_1^2}{4\sigma_1^2\sigma_2^2} z \right) I_0\left(\frac{\sigma_2^2 - \sigma_1^2}{4\sigma_1^2\sigma_2^2} z \right) - 2Q_1\left(\frac{|\sigma_2 - \sigma_1|}{2\sigma_1\sigma_2} \sqrt{z}, \frac{\sigma_2 + \sigma_1}{2\sigma_1\sigma_2} \sqrt{z} \right),$$

$$z \geq 0$$

$$(5.8)$$

$$\Psi_Z(\omega) = \left(\frac{1}{(1 - 2j\omega\sigma_1^2)(1 - 2j\omega\sigma_2^2)} \right)^{1/2}$$

$$(5.9)$$

$$E\{Z^k\} = \frac{2^{2k+1} k! (\sigma_1\sigma_2)^{2k+1}}{(\sigma_1^2 + \sigma_2^2)^{k+1}} \, {}_2F_1\left(\frac{k+1}{2}, \frac{k}{2} + 1; 1; \left(\frac{\sigma_1^2 - \sigma_2^2}{\sigma_1^2 + \sigma_2^2} \right)^2 \right), \quad k \text{ integer} \qquad (5.10)$$

2. $n_1 = n_2 = 2$

$$p_Z(z) = \frac{1}{2(\sigma_2^2 - \sigma_1^2)} \left[\exp\left(-\frac{z}{2\sigma_2^2} \right) - \exp\left(-\frac{z}{2\sigma_1^2} \right) \right], \quad z \geq 0 \qquad (5.11)$$

$$P_Z(z) = 1 - \left(\frac{\sigma_2^2}{\sigma_2^2 - \sigma_1^2} \right) \exp\left(-\frac{z}{2\sigma_2^2} \right) + \left(\frac{\sigma_1^2}{\sigma_2^2 - \sigma_1^2} \right) \exp\left(-\frac{z}{2\sigma_1^2} \right), \quad z \geq 0 \qquad (5.12)$$

$$\Psi_Z(\omega) = \frac{1}{(1 - 2j\omega\sigma_1^2)(1 - 2j\omega\sigma_2^2)} \qquad (5.13)$$

$$E\{Z^k\} = \frac{2^{2k+2}(k+1)!(\sigma_1\sigma_2)^{2k+2}}{(\sigma_1^2+\sigma_2^2)^{k+2}} {}_2F_1\left(\frac{k}{2}+1,\frac{k+3}{2};\frac{3}{2};\left(\frac{\sigma_1^2-\sigma_2^2}{\sigma_1^2+\sigma_2^2}\right)^2\right), \ k \text{ integer}$$

(5.14)

3. $n_1 = n_2 = 2m$

$$p_Z(z) = \frac{1}{2\sigma_1^2}\exp\left(-\frac{z}{2\sigma_1^2}\right)\frac{1}{(m-1)!}\left(\frac{\sigma_1^2}{\sigma_1^2-\sigma_2^2}\right)^m \sum_{i=0}^{m-1}\frac{(2(m-1)-i)!}{i!(m-1-i)!}$$

$$\times\left(\frac{\sigma_2^2}{\sigma_2^2-\sigma_1^2}\right)^{m-1-i}\left(\frac{z}{2\sigma_1^2}\right)^i$$

(5.15)

$$+\frac{1}{2\sigma_2^2}\exp\left(-\frac{z}{2\sigma_2^2}\right)\frac{1}{(m-1)!}\left(\frac{\sigma_2^2}{\sigma_2^2-\sigma_1^2}\right)^m \sum_{i=0}^{m-1}\frac{(2(m-1)-i)!}{i!(m-1-i)!}$$

$$\times\left(\frac{\sigma_1^2}{\sigma_1^2-\sigma_2^2}\right)^{m-1-i}\left(\frac{z}{2\sigma_2^2}\right)^i, \ z \geq 0$$

$$P_Z(z) = 1-\exp\left(-\frac{z}{2\sigma_1^2}\right)\frac{1}{(m-1)!}\left(\frac{\sigma_1^2}{\sigma_1^2-\sigma_2^2}\right)^m \sum_{i=0}^{m-1}\sum_{l=0}^{i}\frac{(2(m-1)-i)!}{(i-l)!(m-1-i)!}$$

$$\times\left(\frac{\sigma_2^2}{\sigma_2^2-\sigma_1^2}\right)^{m-1-i}\left(\frac{z}{2\sigma_1^2}\right)^{i-l}$$

$$-\exp\left(-\frac{z}{2\sigma_2^2}\right)\frac{1}{(m-1)!}\left(\frac{\sigma_2^2}{\sigma_2^2-\sigma_1^2}\right)^m \sum_{i=0}^{m-1}\sum_{l=0}^{i}\frac{(2(m-1)-i)!}{(i-l)!(m-1-i)!}$$

$$\times\left(\frac{\sigma_1^2}{\sigma_1^2-\sigma_2^2}\right)^{m-1-i}\left(\frac{z}{2\sigma_2^2}\right)^{i-l}, \ z \geq 0$$

(5.16)

$$\Psi_Z(\omega) = \left(\frac{1}{(1-2j\omega\sigma_1^2)(1-2j\omega\sigma_2^2)}\right)^m$$

(5.17)

Note that when $\sigma_2^2 = \sigma_1^2 = \sigma^2$, then Z simply becomes a central chi-square RV with $4m$ degrees of freedom with PDF, CDF, and CF determined from (2.32), (2.33), and (2.34), respectively, with m replaced by $2m$. The moments of Z are given by

$$E\{Z^k\} = \frac{2^{2k+2m}(k+2m+1)!(\sigma_1\sigma_2)^{2k+2m}}{(2m-1)!(\sigma_1^2+\sigma_2^2)^{k+2m}}$$

$$\times {}_2F_1\left(m+\frac{k}{2},m+\frac{k+1}{2};m+\frac{1}{2};\left(\frac{\sigma_1^2-\sigma_2^2}{\sigma_1^2+\sigma_2^2}\right)^2\right),\ k\ \text{integer}$$

(5.18)

4. $n_1 = n_2 = 2m+1$

$$p_Z(z) = \frac{\sqrt{\pi}}{2\sigma_1\sigma_2\Gamma(m+1/2)}\left(\frac{z}{2\left|\sigma_2^2-\sigma_1^2\right|}\right)^m \exp\left(-\frac{\sigma_2^2+\sigma_1^2}{4\sigma_1^2\sigma_2^2}z\right)I_m\left(\frac{\sigma_2^2-\sigma_1^2}{4\sigma_1^2\sigma_2^2}z\right),$$ (5.19)

$$z \geq 0$$

$$P_Z(z) = -------$$ (5.20)

$$\Psi_Z(\omega) = \left(\frac{1}{(1-2j\omega\sigma_1^2)(1-2j\omega\sigma_2^2)}\right)^{m+1/2}$$ (5.21)

$$E\{Z^k\} = \frac{2^{2(k+m)+1}(2m+k)!(\sigma_1\sigma_2)^{2(k+m)+1}}{(2m)!(\sigma_1^2+\sigma_2^2)^{2m+k+1}}$$

$$\times {}_2F_1\left(m+\frac{k+1}{2},m+1+\frac{k}{2};m+1;\left(\frac{\sigma_1^2-\sigma_2^2}{\sigma_1^2+\sigma_2^2}\right)^2\right),\ k\ \text{integer}$$

(5.22)

5. $n_1 = 2m, n_2 = 2$

Using (4.13) in (5.4), we immediately arrive at

$$p_Z(z) = \frac{1}{2\sigma_2^2}\left[\left(\frac{\sigma_2^2}{\sigma_2^2-\sigma_1^2}\right)^m \exp\left(-\frac{z}{2\sigma_2^2}\right) - \exp\left(-\frac{z}{2\sigma_1^2}\right)\right]$$

$$\times \sum_{i=0}^{m-1}\frac{1}{i!}\left(\frac{\sigma_2^2}{\sigma_2^2-\sigma_1^2}\right)^{m-i}\left(\frac{z}{2\sigma_1^2}\right)^i\right],\ z \geq 0$$

(5.23)

which, of course, reduces to (5.11) when $m=1$. The corresponding expression for the CDF can be obtained by using (4.14) in (5.6) with the result

$$P_Z(z) = 1 + \exp\left(-\frac{z}{2\sigma_1^2}\right)\left(\frac{\sigma_1^2}{\sigma_2^2 - \sigma_1^2}\right)^{m-1}\sum_{i=0}^{m-1}\sum_{l=0}^{i}\frac{1}{(i-l)!}\left(\frac{\sigma_2^2}{\sigma_2^2 - \sigma_1^2}\right)^{m-1-i}\left(\frac{z}{2\sigma_1^2}\right)^{i-l}$$

$$-\left(\frac{\sigma_2^2}{\sigma_2^2 - \sigma_1^2}\right)^{m}\exp\left(-\frac{z}{2\sigma_2^2}\right), z \geq 0 \tag{5.24}$$

$$\Psi_Z(\omega) = \frac{1}{\left(1 - 2j\omega\sigma_1^2\right)^{m}\left(1 - 2j\omega\sigma_2^2\right)} \tag{5.25}$$

Note that when $\sigma_2^2 = \sigma_1^2 = \sigma^2$, then Z simply becomes a central chi-square RV with $2(m+1)$ degrees of freedom with PDF, CDF and CF determined from (2.32), (2.33), and (2.34), respectively, with m replaced by $m+1$.

6. n_1, n_2

$$P_Z(z) = \frac{1}{2\sigma_1\sigma_2\Gamma\left(\frac{n_1 + n_2}{2}\right)}\left(\frac{z}{2\sigma_1^2}\right)^{(n_1-1)/2}\left(\frac{z}{2\sigma_2^2}\right)^{(n_2-1)/2}\exp\left(-\frac{z}{2\sigma_1^2}\right)$$

$$\times {}_1F_1\left(\frac{n_2}{2}; \frac{n_1 + n_2}{2}; \frac{(\sigma_2^2 - \sigma_1^2)^2}{2\sigma_1^2\sigma_2^2}z\right), z \geq 0 \tag{5.26}$$

$$P_Z(z) = ------ \tag{5.27}$$

$$\Psi_Z(\omega) = \frac{1}{\left(1 - 2j\omega\sigma_1^2\right)^{n_1/2}\left(1 - 2j\omega\sigma_2^2\right)^{n_2/2}} \tag{5.28}$$

B. Dependent Central Chi-Square (+) Central Chi-Square

1. $n_1 = n_2 = 1$

To simplify the expressions, we introduce the parameters

$$\gamma^+ = \frac{\left[(\sigma_2^2 + \sigma_1^2)^2 - 4\sigma_1^2\sigma_2^2(1-\rho^2)\right]^{1/2}}{\sigma_1^2\sigma_2^2(1-\rho^2)}, \quad \beta^{\pm} = \gamma^+ \pm \frac{\sigma_2^2 + \sigma_1^2}{\sigma_1^2\sigma_2^2(1-\rho^2)} \tag{5.29}$$

Note that $\beta^+ \geq 0$ but $\beta^- \leq 0$. Then,

$$p_Z(z) = \frac{1}{2\sigma_1\sigma_2\sqrt{1-\rho^2}} \exp\left(-\frac{1}{4}(\beta^+ - \gamma^+)z\right) I_0\left(\frac{1}{4}\gamma^+ z\right), z \geq 0 \quad (5.30)$$

$$P_Z(z) = 1 + \exp\left(-\frac{1}{4}(\beta^+ - \gamma^+)z\right) I_0\left(\frac{1}{4}\gamma^+ z\right)$$

$$-2Q_1\left(\frac{\sqrt{\sigma_1^2 + \sigma_2^2 - 2\sigma_1\sigma_2\sqrt{1-\rho^2}}}{2\sigma_1\sigma_2\sqrt{1-\rho^2}}\sqrt{z}, \frac{\sqrt{\sigma_1^2 + \sigma_2^2 + 2\sigma_1\sigma_2\sqrt{1-\rho^2}}}{2\sigma_1\sigma_2\sqrt{1-\rho^2}}\sqrt{z}\right),$$

$$z \geq 0$$
$$(5.31)$$

$$\Psi_Z(\omega) = \left(\frac{1-\rho^2}{\left(1-2j\omega(1-\rho^2)\sigma_1^2\right)\left(1-2j\omega(1-\rho^2)\sigma_2^2\right) - \rho^2}\right)^{1/2} \quad (5.32)$$

$$E\{Z^k\} = \frac{2^{2k+1}k!}{\sigma_1\sigma_2\sqrt{1-\rho^2}(\beta^+ - \gamma^+)^{k+1}} \; {}_2F_1\left(\frac{k+1}{2}, \frac{k}{2}+1; 1; \left(\frac{\gamma^+}{\beta^+ - \gamma^+}\right)^2\right),$$

$$k \text{ integer}$$
$$(5.33)$$

2. $n_1 = n_2 = 2$

$$p_Z(z) = \frac{1}{2\sigma_1^2\sigma_2^2(1-\rho^2)\gamma^+}\left[\exp\left(\frac{1}{4}\beta^- z\right) - \exp\left(-\frac{1}{4}\beta^+ z\right)\right], z \geq 0 \quad (5.34)$$

$$P_Z(z) = \frac{2}{\sigma_1^2\sigma_2^2(1-\rho^2)\gamma^+}\left\{-\frac{1}{\beta^-}\left[1 - \exp\left(\frac{1}{4}\beta^- z\right)\right] - \frac{1}{\beta^+}\left[1 - \exp\left(-\frac{1}{4}\beta^+ z\right)\right]\right\},$$

$$z \geq 0$$
$$(5.35)$$

$$\Psi_Z(\omega) = \frac{1-\rho^2}{\left(1-2j\omega(1-\rho^2)\sigma_1^2\right)\left(1-2j\omega(1-\rho^2)\sigma_2^2\right) - \rho^2} \quad (5.36)$$

$$E\{Z^k\} = \frac{2^{2(k+1)}(k+1)!}{\sigma_1^2 \sigma_2^2 (1-\rho^2)(\beta^+ - \gamma^+)^{k+2}} \, {}_2F_1\left(\frac{k}{2}+1, \frac{k+3}{2}; \frac{3}{2}; \left(\frac{\gamma^+}{\beta^+ - \gamma^+}\right)^2\right),$$

$$k \text{ integer}$$

$$(5.37)$$

3. $n_1 = n_2 = 2m$

$$P_Z(z) = \frac{z^{m-1}}{(m-1)!\left[2\sigma_1^2\sigma_2^2(1-\rho^2)\gamma^+\right]^m} \left[\exp\left(\frac{1}{4}\beta^- z\right)\sum_{i=0}^{m-1}\frac{(m+i-1)!}{i!(m-i-1)!}\left(\frac{2}{\gamma^+ z}\right)^i \right.$$

$$\left. +(-1)^m \exp\left(-\frac{1}{4}\beta^+ z\right)\sum_{i=0}^{m-1}\frac{(m+i-1)!}{i!(m-i-1)!}\left(\frac{2}{\gamma^+ z}\right)^i\right], z \ge 0$$

$$(5.38)$$

$$P_Z(z) = \frac{1}{(m-1)!\left(2\sigma_1^2\sigma_2^2(1-\rho^2)\gamma^+\right)^m}\sum_{i=0}^{m-1}\frac{(-1)^i(m+i-1)!}{i!}\left(\frac{2}{\gamma^+}\right)^i$$

$$\times\left\{\left[\frac{1}{(-\beta^-/4)^{m-i}} - \exp\left(\frac{1}{4}\beta^- z\right)\sum_{l=0}^{m-i-1}\frac{1}{(m-i-l-1)!(-\beta^-/4)^{l+1}}z^{m-i-l-1}\right]\right.$$

$$\left. +(-1)^{m-i}\left[\frac{1}{(\beta^+/4)^{m-i}} - \exp\left(-\frac{1}{4}\beta^+ z\right)\sum_{l=0}^{m-i-1}\frac{1}{(m-i-l-1)!(\beta^+/4)^{l+1}}z^{m-i-l-1}\right]\right\},$$

$$z \ge 0$$

$$(5.39)$$

$$\Psi_Z(\omega) = \left(\frac{1-\rho^2}{\left(1-2j\omega(1-\rho^2)\sigma_1^2\right)\left(1-2j\omega(1-\rho^2)\sigma_2^2\right)-\rho^2}\right)^m \qquad (5.40)$$

4. $n_1 = n_2 = 2m+1$

$$P_Z(z) = \frac{\sqrt{\pi/2}}{\left[2\sigma_1^2\sigma_2^2(1-\rho^2)\right]^{m+1/2}\Gamma(m+1/2)}\left(\frac{z}{(\gamma^+)^2}\right)^m \exp\left(-\frac{1}{4}(\beta^+ - \gamma^+)z\right)$$

$$\times I_m\left(\frac{1}{4}\gamma^+ z\right), z \ge 0$$

$$(5.41)$$

$$P_Z(z) = ------\qquad (5.42)$$

$$\Psi_Z(\omega) = \left(\frac{1-\rho^2}{\left(1-2j\omega(1-\rho^2)\sigma_1^2\right)\left(1-2j\omega(1-\rho^2)\sigma_2^2\right)-\rho^2}\right)^{m+1/2}\qquad (5.43)$$

$$E\{Z^k\} = \frac{2^{2(m+k)+1}(2m+k)!}{\left(\sigma_1\sigma_2\sqrt{1-\rho^2}\right)^{2m+1}\left(\beta^+ - \gamma^+\right)^{2m+1+k}}$$

$$\times {}_2F_1\left(m+\frac{k+1}{2}, m+1+\frac{k}{2}; m+1; \left(\frac{\gamma^+}{\beta^+-\gamma^+}\right)^2\right), \; k \text{ integer}$$

(5.44)

C. Independent Noncentral Chi-Square (+) Central Chi-Square

Define $Z = Y_1 + Y_2$ where Y_1 and Y_2 are independent noncentral and central chi-square distributed RVs with n_1 and n_2 degrees of freedom, respectively.

1. $n_1 = n_2 = n$

$$p_Z(z) = \frac{1}{2\sigma_1^2}\left(\frac{\sigma_1}{\sigma_2}\right)^n \left(\frac{z}{a_1^2}\right)^{(n-1)/2} \exp\left(-\frac{z+a_1^2}{2\sigma_1^2}\right)$$

$$\times \sum_{i=0}^{\infty} \frac{\Gamma(n/2+i)}{i!\,\Gamma(n/2)}\left(\frac{\sqrt{z}(\sigma_2^2 - \sigma_1^2)}{a_1\sigma_2^2}\right)^i I_{n+i-1}\left(\frac{\sqrt{z}a_1}{\sigma_1^2}\right), \; z \ge 0$$

(5.45)

$$P_Z(z) = \left(\frac{\sigma_1}{\sigma_2}\right)^n \sum_{i=0}^{\infty} \frac{\Gamma(n/2+i)}{i!\,\Gamma(n/2)}\left(\frac{\sigma_2^2 - \sigma_1^2}{\sigma_2^2}\right)^i \left[1 - Q_{n+i}\left(\frac{a_1}{\sigma_1}, \frac{\sqrt{z}}{\sigma_1}\right)\right], \; z \ge 0 \quad (5.46)$$

$$\Psi_Z(\omega) = \left(\frac{1}{(1-2j\omega\sigma_1^2)(1-2j\omega\sigma_2^2)}\right)^{n/2} \exp\left(\frac{j\omega a_1^2}{1-2j\omega\sigma_1^2}\right) \qquad (5.47)$$

2. $n_1 = 2m_1, n_2 = 2m_2$

$$p_Z(z) = \frac{1}{2\sigma_1^2}\left(\frac{\sigma_1}{\sigma_2}\right)^{2m_2}\left(\frac{z}{a_1^2}\right)^{(m_1+m_2-1)/2}\exp\left(-\frac{z+a_1^2}{2\sigma_1^2}\right)$$

$$\times \sum_{i=0}^{\infty}\frac{\Gamma(m_2+i)}{i!\,\Gamma(m_2)}\left(\frac{\sqrt{z}(\sigma_2^2-\sigma_1^2)}{a_1\sigma_2^2}\right)^i I_{m_1+m_2+i-1}\left(\frac{a_1\sqrt{z}}{\sigma_1^2}\right),\ z \geq 0$$

(5.48)

$$P_Z(z) = \left(\frac{\sigma_1}{\sigma_2}\right)^{2m_2}\sum_{i=0}^{\infty}\frac{\Gamma(m_2+i)}{i!\,\Gamma(m_2)}\left(\frac{\sigma_2^2-\sigma_1^2}{\sigma_2^2}\right)^i\left[1-Q_{m_1+m_2+i}\left(\frac{a_1}{\sigma_1},\frac{\sqrt{z}}{\sigma_1}\right)\right],\ z \geq 0$$

(5.49)

$$\Psi_Z(\omega) = \frac{1}{\left(1-2j\omega\sigma_1^2\right)^{m_1}\left(1-2j\omega\sigma_2^2\right)^{m_2}}\exp\left(\frac{j\omega a_1^2}{1-2j\omega\sigma_1^2}\right)$$

(5.50)

3. $n_1 = 2m, n_2 = 2$

For $\sigma_2^2 > \sigma_1^2$, applying (5.4) and (5.6) to (4.35) and (4.36) gives

$$p_Z(z) = \frac{1}{2\sigma_2^2}\left(\frac{\sigma_2^2}{\sigma_2^2-\sigma_1^2}\right)^m\exp\left(-\frac{z}{2\sigma_2^2}\right)\exp\left(\frac{a_1^2}{2(\sigma_2^2-\sigma_1^2)}\right)$$

$$\times\left[1-Q_m\left(\frac{a_1}{\sigma_1}\sqrt{\frac{\sigma_2^2}{\sigma_2^2-\sigma_1^2}},\sqrt{\frac{z(\sigma_2^2-\sigma_1^2)}{\sigma_1^2\sigma_2^2}}\right)\right],\ z \geq 0$$

(5.51)

$$P_Z(z) = 1 - Q_m\left(\frac{a_1}{\sigma_1},\frac{\sqrt{z}}{\sigma_1}\right) - \left(\frac{\sigma_2^2}{\sigma_2^2-\sigma_1^2}\right)^m\exp\left(-\frac{z}{2\sigma_2^2}\right)\exp\left(\frac{a_1^2}{2(\sigma_2^2-\sigma_1^2)}\right)$$

$$\times\left[1-Q_m\left(\frac{a_1}{\sigma_1}\sqrt{\frac{\sigma_2^2}{\sigma_2^2-\sigma_1^2}},\sqrt{\frac{z(\sigma_2^2-\sigma_1^2)}{\sigma_1^2\sigma_2^2}}\right)\right],\ z \geq 0$$

(5.52)

$$\Psi_Z(\omega) = \frac{1}{\left(1-2j\omega\sigma_1^2\right)^m\left(1-2j\omega\sigma_2^2\right)}\exp\left(\frac{j\omega a_1^2}{1-2j\omega\sigma_1^2}\right)$$

(5.53)

For the limited case of $\sigma_2^2 = \sigma_1^2 = \sigma^2$, one can use the series expansion of the generalized Marcum Q-function, namely,

$$1 - Q_k(\alpha, \beta) = \exp\left(-\frac{\alpha^2 + \beta^2}{2}\right) \sum_{l=k}^{\infty} \left(\frac{\beta}{\alpha}\right)^l I_l(\alpha\beta) \qquad (5.54)$$

in (5.51) to arrive at the results

$$p_Z(z) = \frac{1}{2\sigma^2} \left(\frac{\sqrt{z}}{a_1}\right)^m \exp\left(-\frac{z + a_1^2}{2\sigma^2}\right) I_m\left(\frac{a_1 \sqrt{z}}{\sigma^2}\right), z \ge 0 \qquad (5.55)$$

$$P_Z(z) = 1 - Q_{m+1}\left(\frac{a_1}{\sigma}, \frac{\sqrt{z}}{\sigma}\right), z \ge 0 \qquad (5.56)$$

These results could also have been immediately obtained by noting that, for this limiting case, Z is simply a noncentral chi-square RV with $2(m+1)$ degrees of freedom and the value of a_1 is still obtained from (1.14) since the addition of the central chi-square RV to Y_1 does not change this value.

For $\sigma_2^2 < \sigma_1^2$, the form of the PDF and CDF change with respect to those given in (5.51) and (5.52). Still we can apply the series expansion of the generalized Marcum Q-function to these equations keeping in mind that now the arguments of the Marcum Q-function, α and β, will be purely imaginary. Carrying out the algebra and recalling that $I_l(-x) = (-1)^l I_l(x)$, we obtain

$$p_Z(z) = \frac{1}{2\sigma_2^2} \exp\left(-\frac{z + a_1^2}{2\sigma_1^2}\right) \sum_{l=m}^{\infty} \left(\frac{\sigma_2^2 - \sigma_1^2}{\sigma_2^2}\right)^{l-m} \left(\frac{\sqrt{z}}{a_1}\right)^l I_l\left(\frac{a_1 \sqrt{z}}{\sigma_1^2}\right), z \ge 0 \qquad (5.57)$$

and

$$P_Z(z) = 1 - Q_m\left(\frac{a_1}{\sigma_1}, \frac{\sqrt{z}}{\sigma_1}\right) - \exp\left(-\frac{z + a_1^2}{2\sigma_1^2}\right) \sum_{l=m}^{\infty} \left(\frac{\sigma_2^2 - \sigma_1^2}{\sigma_2^2}\right)^{l-m} \left(\frac{\sqrt{z}}{a_1}\right)^l I_l\left(\frac{a_1 \sqrt{z}}{\sigma_1^2}\right), z \ge 0$$

$$(5.58)$$

which can also be put into the form

$$P_Z(z) = \left(\frac{\sigma_1}{\sigma_2}\right)^2 \sum_{i=0}^{\infty} \left(\frac{\sigma_2^2 - \sigma_1^2}{\sigma_2^2}\right)^i \left[1 - Q_{m_1+1+i}\left(\frac{a_1}{\sigma_1}, \frac{\sqrt{z}}{\sigma_1}\right)\right], z \ge 0 \qquad (5.59)$$

D. Independent Noncentral Chi-Square (+) Noncentral Chi-Square

Define $Z = Y_1 + Y_2$ where Y_1 and Y_2 are independent noncentral chi-square distributed RVs with n_1 and n_2 degrees of freedom, respectively.

1. $n_1 = n_2 = n$

$$p_Z(z) = \frac{1}{2\sigma_1^2}\left(\frac{\sigma_1}{\sigma_2}\right)^n \left(\frac{z}{a_1^2}\right)^{(n-1)/2} \exp\left(-\frac{z}{2\sigma_1^2}\right)\exp\left[-\frac{1}{2}\left(\frac{a_1^2}{\sigma_1^2}+\frac{a_2^2}{\sigma_2^2}\right)\right]$$

$$\times \sum_{i=0}^{\infty}\sum_{l=0}^{\infty}\frac{\Gamma(n/2+i+l)}{i!l!\,\Gamma(n/2+l)}\left(\frac{\sqrt{z}a_2^2\sigma_1^2}{2a_1\sigma_2^4}\right)^l \left(\frac{\sqrt{z}(\sigma_2^2-\sigma_1^2)}{a_1\sigma_2^2}\right)^i I_{n+i+l-1}\left(\frac{\sqrt{z}a_1}{\sigma_1^2}\right), z \geq 0$$

$$\tag{5.60}$$

$$P_Z(z) = \left(\frac{\sigma_1}{\sigma_2}\right)^n \exp\left(-\frac{a_2^2}{2\sigma_2^2}\right)\sum_{i=0}^{\infty}\sum_{l=0}^{\infty}\frac{\Gamma(n/2+i+l)}{i!l!\,\Gamma(n/2+l)}\left(\frac{a_2^2\sigma_1^2}{2\sigma_2^2}\right)^l \left(\frac{\sigma_2^2-\sigma_1^2}{\sigma_2^2}\right)^i$$

$$\times\left[1-Q_{n+i+l}\left(\frac{a_1}{\sigma_1},\frac{\sqrt{z}}{\sigma_1}\right)\right], z \geq 0$$

$$\tag{5.61}$$

$$\Psi_Z(\omega) = \left(\frac{1}{(1-2j\omega\sigma_1^2)(1-2j\omega\sigma_2^2)}\right)^{n/2} \exp\left(\frac{j\omega a_1^2}{1-2j\omega\sigma_1^2}\right)\exp\left(\frac{j\omega a_2^2}{1-2j\omega\sigma_2^2}\right) \tag{5.62}$$

2. $n_1 = 2m_1, n_2 = 2m_2$

$$p_Z(z) = \frac{1}{2\sigma_1^2}\left(\frac{\sigma_1}{\sigma_2}\right)^{2m_2}\left(\frac{z}{a_1^2}\right)^{(m_1+m_2-1)/2}\exp\left(-\frac{z}{2\sigma_1^2}\right)\exp\left[-\frac{1}{2}\left(\frac{a_1^2}{\sigma_1^2}+\frac{a_2^2}{\sigma_2^2}\right)\right]$$

$$\times\sum_{i=0}^{\infty}\sum_{l=0}^{\infty}\frac{\Gamma(m_2+i+l)}{i!l!\,\Gamma(m_2+l)}\left(\frac{\sqrt{z}a_2^2\sigma_1^2}{2a_1\sigma_2^4}\right)^l \left(\frac{\sqrt{z}(\sigma_2^2-\sigma_1^2)}{a_1\sigma_2^2}\right)^i I_{m_1+m_2+i+l-1}\left(\frac{\sqrt{z}a_1}{\sigma_1^2}\right),$$

$$z \geq 0 \tag{5.63}$$

$$P_Z(z) = \left(\frac{\sigma_1}{\sigma_2}\right)^{2m_2} \exp\left(-\frac{a_2^2}{2\sigma_2^2}\right) \sum_{i=0}^{\infty} \sum_{l=0}^{\infty} \frac{\Gamma(m_2+i+l)}{i!l!\Gamma(m_2+l)} \left(\frac{a_2^2\sigma_1^2}{2\sigma_2^2}\right)^l \left(\frac{\sigma_2^2-\sigma_1^2}{\sigma_2^2}\right)^i$$

$$\times \left[1 - Q_{m_1+m_2+i+l}\left(\frac{a_1}{\sigma_1}, \frac{\sqrt{z}}{\sigma_1}\right)\right], z \geq 0 \tag{5.64}$$

$$\Psi_Z(\omega) = \frac{1}{\left(1-2j\omega\sigma_1^2\right)^{m_1}\left(1-2j\omega\sigma_2^2\right)^{m_2}} \exp\left(\frac{j\omega a_1^2}{1-2j\omega\sigma_1^2}\right) \exp\left(\frac{j\omega a_2^2}{1-2j\omega\sigma_2^2}\right) \tag{5.65}$$

6

PRODUCTS OF RANDOM VARIABLES[4]

A. Independent Gaussian (×) Gaussian (Both Have Zero Mean)

Let $\mathbf{X}^{(1)} \in N_n(\mathbf{0}, \sigma_1^2)$ and $\mathbf{X}^{(2)} \in N_n(\mathbf{0}, \sigma_2^2)$ be independent Gaussian vectors. Then, the inner product of these vectors, namely,

$$X = \left(\mathbf{X}^{(1)}, \mathbf{X}^{(2)}\right) = \sum_{k=1}^{n} X_k^{(1)} X_k^{(2)} \tag{6.1}$$

has the following statistical properties.

1. $n = 1$

$$p_X(x) = \frac{1}{\pi \sigma_1 \sigma_2} K_0\left(\frac{|x|}{\sigma_1 \sigma_2}\right) \tag{6.2}$$

$$P_X(x) = - - - - - - \tag{6.3}$$

$$\Psi_X(\omega) = \left(\frac{1}{1 + \sigma_1^2 \sigma_2^2 \omega^2}\right)^{1/2} \tag{6.4}$$

2. $n = 2$

---

[4] A large number of the PDF and statistical moment results in this section come from Ref. 5.

$$p_X(x) = \frac{1}{2\sigma_1\sigma_2} \exp\left(-\frac{|x|}{\sigma_1\sigma_2}\right) \tag{6.5}$$

$$P_X(x) = \begin{cases} \dfrac{1}{2}\exp\left(\dfrac{x}{\sigma_1\sigma_2}\right), & x < 0 \\[3mm] 1 - \dfrac{1}{2}\exp\left(-\dfrac{x}{\sigma_1\sigma_2}\right), & x \geq 0 \end{cases} \tag{6.6}$$

$$\Psi_X(\omega) = \frac{1}{1+\sigma_1^2\sigma_2^2\omega^2} \tag{6.7}$$

$$E\{X^k\} = \begin{cases} \left(\dfrac{1}{2}\sigma_1\sigma_2\right)^k \displaystyle\sum_{\substack{i=0 \\ i\ \text{even}}}^{k} \binom{k}{i}\left[\dfrac{i!}{(i/2)!}\right]^2\left[\dfrac{(k-i)!}{((k-i)/2)!}\right]^2, & k\ \text{even} \\[3mm] 0, & k\ \text{odd} \end{cases} \tag{6.8}$$

3. $n = 2m$

$$p_X(x) = \frac{1}{\sigma_1\sigma_2(m-1)!}\exp\left(-\frac{|x|}{\sigma_1\sigma_2}\right)\sum_{i=0}^{m-1}\frac{(m+i-1)!}{2^{m+i}i!(m-i-1)!}\left(\frac{|x|}{\sigma_1\sigma_2}\right)^{m-1-i} \tag{6.9}$$

$$P_X(x) = \begin{cases} \dfrac{1}{(m-1)!}\exp\left(\dfrac{x}{\sigma_1\sigma_2}\right)\displaystyle\sum_{i=0}^{m-1}\sum_{l=0}^{m-i-1}\dfrac{(m+i-1)!}{2^{m+i}i!(m-i-1-l)!}\left(-\dfrac{x}{\sigma_1\sigma_2}\right)^{m-1-i-l}, \\[2mm] \hspace{9cm} x < 0 \\[4mm] 1 - \dfrac{1}{(m-1)!}\exp\left(-\dfrac{x}{\sigma_1\sigma_2}\right)\displaystyle\sum_{i=0}^{m-1}\sum_{l=0}^{m-i-1}\dfrac{(m+i-1)!}{2^{m+i}i!(m-i-1-l)!}\left(\dfrac{x}{\sigma_1\sigma_2}\right)^{m-1-i-l}, \\[2mm] \hspace{9cm} x \geq 0 \end{cases} \tag{6.10}$$

$$\Psi_X(\omega) = \left(\frac{1}{1+\sigma_1^2\sigma_2^2\omega^2}\right)^m \tag{6.11}$$

4. $n = 2m+1$

$$p_X(x) = \frac{\left(|x|/2\sigma_1\sigma_2\right)^m}{\sqrt{\pi}\Gamma(m+1/2)\sigma_1\sigma_2} K_m\left(\frac{|x|}{\sigma_1\sigma_2}\right)$$

(6.12)

$$P_X(x) = - - - - - -$$

(6.13)

$$\Psi_X(\omega) = \left(\frac{1}{1+\sigma_1^2\sigma_2^2\omega^2}\right)^{m+1/2}$$

(6.14)

B. Dependent Gaussian (×) Gaussian (Both Have Zero Mean)

Let $X^{(1)} \in N_n(0,\sigma_1^2)$ and $X^{(2)} \in N_n(0,\sigma_2^2)$ be dependent Gaussian vectors. Then, the inner product of these vectors as defined in (6.1) has the following statistical properties.

1. $n = 1$

$$p_X(x) = \frac{1}{\pi\sigma_1\sigma_2} \exp\left(\frac{\rho x}{\sigma_1\sigma_2(1-\rho^2)}\right) K_0\left(\frac{|x|}{\sigma_1\sigma_2(1-\rho^2)}\right)$$

(6.15)

$$P_X(x) = - - - - - -$$

(6.16)

$$\Psi_X(\omega) = \left(\frac{1}{1-2j\omega\rho\sigma_1\sigma_2 + \sigma_1^2\sigma_2^2\omega^2(1-\rho^2)}\right)^{1/2}$$

(6.17)

2. $n = 2$

$$p_X(x) = \frac{1}{2\sigma_1\sigma_2} \exp\left(-\frac{|x|-\rho x}{\sigma_1\sigma_2(1-\rho^2)}\right)$$

(6.18)

$$P_X(x) = \begin{cases} \dfrac{1-\rho}{2}\exp\left(\dfrac{x}{\sigma_1\sigma_2(1-\rho)}\right), & x < 0 \\[4mm] 1 - \dfrac{1+\rho}{2}\exp\left(-\dfrac{x}{\sigma_1\sigma_2(1+\rho)}\right), & x \geq 0 \end{cases}$$

(6.19)

$$\Psi_X(\omega) = \frac{1}{1 - 2j\omega\rho\sigma_1\sigma_2 + \sigma_1^2\sigma_2^2\omega^2(1-\rho^2)} \tag{6.20}$$

$$E\{X^k\} = \sigma^{2k} \sum_{i=0}^{k} \binom{k}{i} \sum_{l=0}^{\lfloor i/2 \rfloor} \sum_{r=0}^{\lfloor (k-i)/2 \rfloor} \binom{i}{2l}\binom{k-i}{2r}(1-\rho^2)^{l+r}\rho^{k-2(l+r)}$$
$$\times (2(i-l)-1)!!(2(k-i-r)-1)!!(2l-1)!!(2r-1)!! \tag{6.21}$$

3. $n = 2m$

$$p_X(x) = \frac{1}{\sigma_1\sigma_2(m-1)!}\exp\left(-\frac{|x|-\rho x}{\sigma_1\sigma_2(1-\rho^2)}\right)\sum_{i=0}^{m-1}\frac{(m+i-1)!}{2^{m+i}\,i!(m-i-1)!}$$
$$\times\left(\frac{|x|}{\sigma_1\sigma_2(1-\rho^2)}\right)^{m-1-i} \tag{6.22}$$

$$P_X(x) = \begin{cases} \dfrac{(1-\rho)^m}{(m-1)!}\exp\left(\dfrac{x}{\sigma_1\sigma_2(1-\rho)}\right)\displaystyle\sum_{i=0}^{m-1}\sum_{l=0}^{m-i-1}\dfrac{(m+i-1)!}{2^{m+i}\,i!(m-i-1-l)!} \\[2mm] \times(1+\rho)^i\left(-\dfrac{x}{\sigma_1\sigma_2(1-\rho)}\right)^{m-1-i-l}, \quad x < 0 \\[4mm] 1 - \dfrac{(1+\rho)^m}{(m-1)!}\exp\left(-\dfrac{x}{\sigma_1\sigma_2(1+\rho)}\right)\displaystyle\sum_{i=0}^{m-1}\sum_{l=0}^{m-i-1}\dfrac{(m+i-1)!}{2^{m+i}\,i!(m-i-1-l)!} \\[2mm] \times(1-\rho)^i\left(\dfrac{x}{\sigma_1\sigma_2(1+\rho)}\right)^{m-1-i-l}, \quad x \geq 0 \end{cases} \tag{6.23}$$

$$\Psi_X(\omega) = \left(\frac{1}{1 - 2j\omega\rho\sigma_1\sigma_2 + \sigma_1^2\sigma_2^2\omega^2(1-\rho^2)}\right)^m \tag{6.24}$$

4. $n = 2m+1$

$$p_X(x) = \frac{(|x|/2\sigma_1\sigma_2)^m}{\sqrt{\pi}\,\Gamma(m+1/2)\sigma_1\sigma_2\sqrt{1-\rho^2}}\exp\left(\frac{\rho x}{\sigma_1\sigma_2(1-\rho^2)}\right)$$
$$\times K_m\left(\frac{|x|}{\sigma_1\sigma_2(1-\rho^2)}\right) \tag{6.25}$$

$$P_X(x) = ------$$ (6.26)

$$\Psi_x(\omega) = \left(\frac{1}{1 - 2j\omega\rho\sigma_1\sigma_2 + \sigma_1^2\sigma_2^2\omega^2(1-\rho^2)} \right)^{m+1/2}$$ (6.27)

C. Independent Gaussian (×) Gaussian (One Has Zero Mean, Both Have Identical Variance)

Let $X^{(1)} \in N_n(\overline{X}^{(1)}, \sigma^2)$ and $X^{(2)} \in N_n(0, \sigma^2)$ be independent Gaussian vectors. Then, the inner product of these vectors as defined in (6.1) has the following statistical properties.

1. $n = 1$

$$P_X(x) = \frac{1}{\sqrt{\pi\sigma^2}} \exp\left(-\frac{a_1^2}{2\sigma^2} \right) \sum_{i=0}^{\infty} \frac{1}{i!\,\Gamma(i+1/2)} \left(\frac{a_1^2|x|}{4\sigma^4} \right)^i K_i\left(\frac{|x|}{\sigma^2} \right)$$ (6.28)

$$P_X(x) = ------$$ (6.29)

$$\Psi_x(\omega) = \left(\frac{1}{1+\sigma^4\omega^2} \right)^{1/2} \exp\left(-\frac{\sigma^2\omega^2 a_1^2}{2(1+\sigma^4\omega^2)} \right)$$ (6.30)

2. $n = 2$

$$P_X(x) = \frac{1}{2\sigma^2} \exp\left(-\frac{|x| + a_1^2/2}{\sigma^2} \right) \sum_{i=0}^{\infty} \sum_{l=0}^{i} \frac{(i+l)!}{2^l(i!)^2 l!(i-l)!} \left(\frac{a_1^2}{4\sigma^2} \right)^i \left(\frac{|x|}{\sigma^2} \right)^{i-l}$$ (6.31)

$$P_X(x) = \frac{1}{2} \exp\left(\frac{x - a_1^2/2}{\sigma^2} \right) \sum_{i=0}^{\infty} \sum_{l=0}^{i} \sum_{r=0}^{i-l} \frac{(i+l)!}{2^l(i!)^2 l!(i-l-r)!} \left(\frac{a_1^2}{4\sigma^2} \right)^i$$
$$\times \left(-\frac{x}{\sigma^2} \right)^{i-l-r}, \quad x < 0$$

$$P_X(x) = 1 - \frac{1}{2}\exp\left(-\frac{x + a_1^2/2}{\sigma^2}\right)\sum_{i=0}^{\infty}\sum_{l=0}^{i}\sum_{r=0}^{i-l}\frac{(i+l)!}{2^l(i!)^2 l!(i-l-r)!}\left(\frac{a_1^2}{4\sigma^2}\right)^i$$

$$\times\left(\frac{x}{\sigma^2}\right)^{i-l-r}, x \geq 0 \tag{6.32}$$

$$\Psi_X(\omega) = \frac{1}{1+\sigma^4\omega^2}\exp\left(-\frac{\sigma^2\omega^2 a_1^2}{2(1+\sigma^4\omega^2)}\right) \tag{6.33}$$

3. $n = 2m$

$$p_X(x) = \frac{1}{2\sigma^2}\left(\frac{|x|}{2\sigma^2}\right)^{m-l}\exp\left(-\frac{|x| + a_1^2/2}{\sigma^2}\right)\sum_{i=0}^{\infty}\sum_{l=0}^{m+i-1}\frac{(m+i+l-1)!}{2^l i!(m+i-1)!}$$

$$\times\frac{1}{l!(m+i-l-1)!}\left(\frac{a_1^2}{4\sigma^2}\right)^i\left(\frac{|x|}{\sigma^2}\right)^{i-l} \tag{6.34}$$

$$P_X(x) = \begin{cases} \dfrac{1}{2^k}\exp\left(\dfrac{x - a_1^2/2}{\sigma^2}\right)\displaystyle\sum_{i=0}^{\infty}\sum_{l=0}^{m+i-1}\sum_{r=0}^{m+i-l-1}\dfrac{(m+i+l-1)!}{2^l i!(m+i-1)!} \\[2mm] \times\dfrac{1}{l!(m+i-l-r-1)!}\left(\dfrac{a_1^2}{4\sigma^2}\right)^i\left(-\dfrac{x}{\sigma^2}\right)^{m+i-l-r-1}, x < 0 \\[4mm] 1 - \dfrac{1}{2^k}\exp\left(-\dfrac{x + a_1^2/2}{\sigma^2}\right)\displaystyle\sum_{i=0}^{\infty}\sum_{l=0}^{m+i-1}\sum_{r=0}^{m+i-l-1}\dfrac{(m+i+l-1)!}{2^l i!(m+i-1)!} \\[2mm] \times\dfrac{1}{l!(m+i-l-r-1)!}\left(\dfrac{a_1^2}{4\sigma^2}\right)^i\left(\dfrac{x}{\sigma^2}\right)^{m+i-l-r-1}, x \geq 0 \end{cases} \tag{6.35}$$

$$\Psi_X(\omega) = \left(\frac{1}{1+\sigma^4\omega^2}\right)^m\exp\left(-\frac{\omega^2\sigma^2 a_1^2}{2(1+\sigma^4\omega^2)}\right) \tag{6.36}$$

4. $n = 2m + 1$

$$p_X(x) = \frac{1}{\sqrt{\pi}\sigma^2}\left(\frac{x}{2\sigma^2}\right)^m\exp\left(-\frac{a_1^2}{2\sigma^2}\right)\sum_{i=0}^{\infty}\frac{1}{i!\Gamma(m+i+1/2)}$$

$$\times\left(\frac{a_1^2|x|}{4\sigma^4}\right)^{m+i}K_{m+i}\left(\frac{|x|}{\sigma^2}\right) \tag{6.37}$$

$$P_X(x) = -\; -\; -\; -\; -\; - \tag{6.38}$$

$$\Psi_X(\omega) = \left(\frac{1}{1+\sigma^4\omega^2}\right)^{m+1/2} \exp\left(-\frac{\sigma^2\omega^2 a_1^2}{2(1+\sigma^4\omega^2)}\right) \tag{6.39}$$

D. Independent Gaussian (×) Gaussian (Both Have Nonzero Mean and Identical Variance)

Let $X^{(1)} \in N_n(\overline{X}^{(1)}, \sigma^2)$ and $X^{(2)} \in N_n(\overline{X}^{(2)}, \sigma^2)$ be independent Gaussian vectors. For this case, the results for the PDF of the inner product of these vectors as defined in (6.1) can be expressed in the form of a doubly infinite series in Whittaker functions. Thus, once again, because of the complexity of the resulting expressions and their somewhat limited use in practical applications, the author has decided to omit these results for this case. Nevertheless, the CF is still relatively simple and given by

$$\Psi_X(\omega) = \left(\frac{1}{1+\sigma^4\omega^2}\right)^{n/2} \exp\left(-\frac{\sigma^2\omega^2\left(a_1^2 + a_2^2\right) - 2j\omega\sum_{l=1}^{n}\overline{X}_l^{(1)}\overline{X}_l^{(2)}}{2(1+\sigma^4\omega^2)}\right) \tag{6.40}$$

A similar statement can be made for the corresponding dependent case. Here the CF is given by

$$\Psi_X(\omega) = \left(\frac{1}{1+\sigma^4\omega^2}\right)^{n/2}$$
$$\times \exp\left(-\frac{\sigma^2\omega^2\left(a_1^2 + a_2^2\right) - 2\left(j\omega + \rho\sigma^2\omega^2\right)\sum_{l=1}^{n}\overline{X}_l^{(1)}\overline{X}_l^{(2)}}{2(1+\sigma^4\omega^2 - 2j\rho\sigma^2\omega)}\right) \tag{6.41}$$

E. Independent Rayleigh (×) Rayleigh

Let $R_1 = \|\mathbf{X}^{(1)}\|$ and $R_2 = \|\mathbf{X}^{(2)}\|$ be Rayleigh RVs corresponding to independent Gaussian vectors $\mathbf{X}^{(1)} \in N_{n_1}(0, \sigma_1^2)$ and $\mathbf{X}^{(2)} \in N_{n_2}(0, \sigma_2^2)$. Then the product RV $R = R_1 R_2$ has the following statistical properties.

1. $n_1 = n_2 = 1$

$$p_R(r) = \frac{2}{\pi \sigma_1 \sigma_2} K_0\left(\frac{r}{\sigma_1 \sigma_2}\right), r \geq 0 \tag{6.42}$$

$$P_R(r) = \text{------} \tag{6.43}$$

$$E\{R^k\} = (2\sigma_1 \sigma_2)^k \frac{\Gamma^2\left(\dfrac{k+1}{2}\right)}{\pi}, k \text{ integer} \tag{6.44}$$

2. $n_1 = n_2 = 2$

$$p_R(r) = \frac{r}{\sigma_1^2 \sigma_2^2} K_0\left(\frac{r}{\sigma_1 \sigma_2}\right), r \geq 0 \tag{6.45}$$

$$P_R(r) = 1 - \frac{r}{\sigma_1 \sigma_2} K_1\left(\frac{r}{\sigma_1 \sigma_2}\right), r \geq 0 \tag{6.46}$$

$$E\{R^k\} = (2\sigma_1 \sigma_2)^k \Gamma^2\left(\frac{k}{2} + 1\right), k \text{ integer} \tag{6.47}$$

3. $n_1 = n_2 = 2m$

$$p_R(r) = \frac{4}{r[(m-1)!]^2}\left(\frac{r}{2\sigma_1 \sigma_2}\right)^{2m} K_0\left(\frac{r}{\sigma_1 \sigma_2}\right), r \geq 0 \tag{6.48}$$

$$P_R(r) = \text{------} \tag{6.49}$$

$$E\{R^k\} = (2\sigma_1 \sigma_2)^k \frac{\Gamma^2\left(\dfrac{k}{2} + m\right)}{[(m-1)!]^2}, k \text{ integer} \tag{6.50}$$

4. n_1, n_2

$$p_R(r) = \frac{4}{r\Gamma(n_1/2)\Gamma(n_2/2)}\left(\frac{r}{2\sigma_1\sigma_2}\right)^{(n_1+n_2)/2} K_{(n_1-n_2)/2}\left(\frac{r}{\sigma_1\sigma_2}\right), r \geq 0 \qquad (6.51)$$

$$P_R(r) = ------ \qquad (6.52)$$

$$E\{R^k\} = (2\sigma_1\sigma_2)^k \frac{\Gamma\left(\dfrac{k+n_1}{2}\right)\Gamma\left(\dfrac{k+n_2}{2}\right)}{\Gamma\left(\dfrac{n_1}{2}\right)\Gamma\left(\dfrac{n_2}{2}\right)}, k \text{ integer} \qquad (6.53)$$

F. Dependent Rayleigh (×) Rayleigh

Let $R_1 = \|\mathbf{X}^{(1)}\|$ and $R_2 = \|\mathbf{X}^{(2)}\|$ be Rayleigh RVs corresponding to dependent Gaussian vectors $\mathbf{X}^{(1)} \in N_n(0, \sigma_1^2)$ and $\mathbf{X}^{(2)} \in N_n(0, \sigma_2^2)$. Then the product RV $R = R_1 R_2$ has the following statistical properties.

$$p_R(r) = \frac{1}{\sigma_1\sigma_2}\left(\frac{r}{\sigma_1\sigma_2}\right)^{n/2}\left(\frac{1-\rho^2}{2|\rho|}\right)^{n/2-1} I_{n/2-1}\left(\frac{r|\rho|}{\sigma_1\sigma_2(1-\rho^2)}\right)$$

$$\times K_0\left(\frac{r}{\sigma_1\sigma_2(1-\rho^2)}\right), r \geq 0 \qquad (6.54)$$

$$p_R(r) = \frac{1}{\sigma_1\sigma_2}\left(\frac{r}{\sigma_1\sigma_2}\right)^{n/2}\left(\frac{1-\rho^2}{2|\rho|}\right)^{n/2-1} I_{n/2-1}\left(\frac{r|\rho|}{\sigma_1\sigma_2(1-\rho^2)}\right)$$

$$\times K_0\left(\frac{r}{\sigma_1\sigma_2(1-\rho^2)}\right), r \geq 0 \qquad (6.55)$$

$$P_R(r) = ------ \qquad (6.56)$$

$$E\{R^k\} = (2\sigma_1\sigma_2)^k (1-\rho^2)^{n+k} \frac{\Gamma^2\left(\dfrac{k+n}{2}\right)}{\Gamma^2\left(\dfrac{n}{2}\right)}$$

$$\times\, {}_2F_1\left(\frac{n+k}{2},\frac{n+k}{2};\frac{n}{2};\rho^2\right), \; k \text{ integer} \tag{6.57}$$

G. Independent Rice (×) Rayleigh

Let $R_1 = \|\mathbf{X}^{(1)}\|$ and $R_2 = \|\mathbf{X}^{(2)}\|$ be Rice and Rayleigh RVs corresponding to independent Gaussian vectors $\mathbf{X}^{(1)} \in N_{n_1}(\overline{\mathbf{X}}^{(1)},\sigma_1^2)$ and $\mathbf{X}^{(2)} \in N_{n_2}(0,\sigma_2^2)$. Then the product RV $R = R_1R_2$ has the following statistical properties.

1. $n_1 = n_2 = 1$

$$p_R(r) = \frac{2}{\pi\sigma_1\sigma_2}\exp\left(-\frac{a_1^2}{2\sigma_1^2}\right)\sum_{i=0}^{\infty}\frac{1}{i!\,\Gamma(i+1/2)}\left(\frac{a_1\sqrt{r}}{2\sigma_1^2}\right)^{2i}\left(\frac{\sigma_1}{\sigma_2}\right)^i K_i\left(\frac{r}{\sigma_1\sigma_2}\right), \; r\geq 0 \tag{6.58}$$

$$P_R(r) = ------- \tag{6.59}$$

2. $n_1 = n_2 = 2$

$$p_R(r) = \frac{r}{\sigma_1^2\sigma_2^2}\exp\left(-\frac{a_1^2}{2\sigma_1^2}\right)\sum_{i=0}^{\infty}\frac{1}{(i!)^2}\left(\frac{a_1\sqrt{r}}{2\sigma_1^2}\right)^{2i}\left(\frac{\sigma_1}{\sigma_2}\right)^i K_i\left(\frac{r}{\sigma_1\sigma_2}\right), \; r\geq 0 \tag{6.60}$$

$$P_R(r) = 1-\exp\left(-\frac{a_1^2}{2\sigma_1^2}\right)\sum_{i=0}^{\infty}\frac{1}{(i!)^2}\left(\frac{a_1}{2\sigma_1}\right)^{2i}\left(\frac{r}{\sigma_1\sigma_2}\right)^{i+1} K_{i+1}\left(\frac{r}{\sigma_1\sigma_2}\right), \; r\geq 0 \tag{6.61}$$

$$E\{R^k\} = (2\sigma_1\sigma_2)^k \Gamma^2\left(\frac{k}{2}+1\right)\exp\left(-\frac{a_1^2}{2\sigma_1^2}\right) {}_1F_1\left(\frac{k}{2}+1;1;\frac{a_1^2}{2\sigma_1^2}\right), \; k \text{ integer} \tag{6.62}$$

3. n_1, n_2

$$p_R(r) = \frac{4}{r\Gamma(n_2/2)} \left(\frac{r}{2\sigma_1\sigma_2}\right)^{(n_1+n_2)/2} \exp\left(-\frac{a_1^2}{2\sigma_1^2}\right)$$

$$\times \sum_{i=0}^{\infty} \frac{1}{i!\Gamma(i+n_2/2)} \left(\frac{a_1\sqrt{r}}{2\sigma_1^2}\right)^{2i} \left(\frac{\sigma_1}{\sigma_2}\right)^i K_{i+(n_1-n_2)/2}\left(\frac{r}{\sigma_1\sigma_2}\right), r \geq 0 \qquad (6.63)$$

$$p_R(r) = ------ \qquad (6.64)$$

$$E\{R^k\} = (2\sigma_1\sigma_2)^k \frac{\Gamma\left(\dfrac{k+n_1}{2}\right)\Gamma\left(\dfrac{k+n_2}{2}\right)}{\Gamma\left(\dfrac{n_1}{2}\right)\Gamma\left(\dfrac{n_2}{2}\right)} \exp\left(-\frac{a_1^2}{2\sigma_1^2}\right) {}_1F_1\left(\frac{k+n_1}{2};\frac{n_1}{2};\frac{a_1^2}{2\sigma_1^2}\right), \qquad (6.65)$$

$$k \text{ integer}$$

H. Independent Rice (×) Rice

Let $R_1 = \|\mathbf{X}^{(1)}\|$ and $R_2 = \|\mathbf{X}^{(2)}\|$ be Rician RVs corresponding to independent Gaussian vectors $\mathbf{X}^{(1)} \in N_{n_1}\left(\overline{\mathbf{X}}^{(1)}, \sigma_1^2\right)$ and $\mathbf{X}^{(2)} \in N_{n_2}\left(\overline{\mathbf{X}}^{(2)}, \sigma_2^2\right)$. Then the product RV $R = R_1 R_2$ has the following statistical properties.

1. n_1, n_2

$$p_R(r) = \frac{4}{r}\left(\frac{r}{2\sigma_1\sigma_2}\right)^{(n_1+n_2)/2} \exp\left[-\frac{1}{2}\left(\frac{a_1^2}{\sigma_1^2}+\frac{a_2^2}{\sigma_2^2}\right)\right]$$

$$\times \sum_{i=0}^{\infty}\sum_{l=0}^{\infty} \frac{1}{i!l!\Gamma(i+n_1/2)\Gamma(l+n_2/2)} \left(\frac{a_1\sqrt{r}}{2\sigma_1^2}\right)^{2i}\left(\frac{a_2\sqrt{r}}{2\sigma_2^2}\right)^{2i} \qquad (6.66)$$

$$\times \left(\frac{\sigma_1}{\sigma_2}\right)^{i-l} K_{i-l+(n_1-n_2)/2}\left(\frac{r}{\sigma_1\sigma_2}\right), r \geq 0$$

$$p_R(r) = ------ \qquad (6.67)$$

$$E\{R^k\} = (2\sigma_1\sigma_2)^k \frac{\Gamma\left(\dfrac{k+n_1}{2}\right)\Gamma\left(\dfrac{k+n_2}{2}\right)}{\Gamma\left(\dfrac{n_1}{2}\right)\Gamma\left(\dfrac{n_2}{2}\right)} \exp\left[-\frac{1}{2}\left(\frac{a_1^2}{\sigma_1^2}+\frac{a_2^2}{\sigma_2^2}\right)\right]$$

$$\times {}_1F_1\left(\frac{k+n_1}{2};\frac{n_1}{2};\frac{a_1^2}{2\sigma_1^2}\right){}_1F_1\left(\frac{k+n_2}{2};\frac{n_2}{2};\frac{a_2^2}{2\sigma_2^2}\right), \quad k \text{ integer}$$

(6.68)

I. Dependent Rayleigh Products

Consider the product RV $U_1 = R_1 R_3$ where $R_1 = \left\|\mathbf{X}^{(1)}\right\|$ and $R_3 = \left\|\mathbf{X}^{(3)}\right\|$ are Rayleigh RVs corresponding to independent Gaussian vectors $\mathbf{X}^{(1)} \in N_n(\mathbf{0}, \sigma_1^2)$ and $\mathbf{X}^{(3)} \in N_n(\mathbf{0}, \sigma_1^2)$. Likewise consider the product RV $U_2 = R_2 R_4$ where $R_2 = \left\|\mathbf{X}^{(2)}\right\|$ and $R_4 = \left\|\mathbf{X}^{(4)}\right\|$ are Rayleigh RVs corresponding to independent Gaussian vectors $\mathbf{X}^{(2)} \in N_n(\mathbf{0}, \sigma_2^2)$ and $\mathbf{X}^{(4)} \in N_n(\mathbf{0}, \sigma_2^2)$. The Gaussian vectors $\mathbf{X}^{(1)}$ and $\mathbf{X}^{(2)}$ are dependent and likewise the Gaussian vectors $\mathbf{X}^{(3)}$ and $\mathbf{X}^{(4)}$ are dependent with correlation matrix as in (1.10). Then U_1 and U_2 have the joint PDF.

$$p_{U_1,U_2}(u_1,u_2) = \frac{1}{2^{2n-4}\sigma_1^2\sigma_2^2\Gamma^2(n/2)}\left(\frac{u_1 u_2}{\sigma_1^2\sigma_2^2}\right)^{n-1}$$

$$\times \sum_{i=0}^{\infty}\sum_{l=0}^{\infty}\frac{1}{i!\,l!\,\Gamma(i+n/2)\Gamma(l+n/2)}\left(\frac{u_1 u_2 \rho^2}{4\sigma_1^2\sigma_2^2(1-\rho^2)^2}\right)^{i+l}$$

(6.69)

$$\times K_{i-l}\left(\frac{u_1}{\sigma_1^2(1-\rho^2)}\right)K_{i-l}\left(\frac{u_2}{\sigma_2^2(1-\rho^2)}\right), \quad u_1, u_2 \geq 0$$

7

RATIOS OF RANDOM VARIABLES[5]

A. Independent Gaussian (÷) Gaussian (Both Have Zero Mean)

Let $X_1 \in N_1(0,\sigma_1^2)$ and $X_2 \in N_1(0,\sigma_2^2)$ be independent Gaussian RVs. Then, the ratio of these RVs $X = X_2 / X_1$ has the following statistical properties.

$$p_X(x) = \frac{\sigma_1 \sigma_2}{\pi(\sigma_1^2 x^2 + \sigma_2^2)} \tag{7.1}$$

$$P_X(x) = \frac{1}{2} + \frac{1}{\pi} \tan^{-1}\left(\frac{\sigma_1 x}{\sigma_2}\right) \tag{7.2}$$

$$\Psi_X(\omega) = \exp\left(-\frac{\sigma_2}{\sigma_1}|\omega|\right) \tag{7.3}$$

B. Independent Gaussian (÷) Gaussian (One Has Zero Mean)

Let $X_1 \in N_1(\overline{X}_1,\sigma_1^2)$ and $X_2 \in N_1(0,\sigma_2^2)$ be independent Gaussian RVs.

---

[5] A large number of the CDF results in this section come from Ref. 11 where they appear in their normalized (the RVs that form the ratio have unit variance) form. Once again, a large number of the PDF and statistical moment results come from Ref. 5.

Then, the ratio of these RVs $X = X_2 / X_1$ has the following statistical properties

$$p_X(x) = \frac{\sigma_1 \sigma_2}{\pi(\sigma_1^2 x^2 + \sigma_2^2)} \exp\left(-\frac{\overline{X}_1^2}{2\sigma_1^2}\right) + \frac{\overline{X}_1 \sigma_2^2}{\sqrt{2\pi}(\sigma_1^2 x^2 + \sigma_2^2)^{3/2}}$$

$$\times \exp\left(-\frac{\overline{X}_1^2 x^2}{2(\sigma_1^2 x^2 + \sigma_2^2)}\right)\left[1 - 2Q\left(\frac{\overline{X}_1 \sigma_2^2}{\sigma_1 \sigma_2 (\sigma_1^2 x^2 + \sigma_2^2)^{1/2}}\right)\right]$$

(7.4)

$$P_X(x) = \begin{cases} \dfrac{1}{\pi} \tan^{-1}\left(\dfrac{\sigma_2}{\sigma_1 |x|}\right) - 2V\left(\dfrac{\overline{X}_1 |x|}{\sqrt{\sigma_1^2 x^2 + \sigma_2^2}}, \dfrac{\overline{X}_1(\sigma_2/\sigma_1)}{\sqrt{\sigma_1^2 x^2 + \sigma_2^2}}\right), & x < 0 \\[4mm] 1 - \dfrac{1}{\pi} \tan^{-1}\left(\dfrac{\sigma_2}{\sigma_1 x}\right) + 2V\left(\dfrac{\overline{X}_1 x}{\sqrt{\sigma_1^2 x^2 + \sigma_2^2}}, \dfrac{\overline{X}_1(\sigma_2/\sigma_1)}{\sqrt{\sigma_1^2 x^2 + \sigma_2^2}}\right), & x \geq 0 \end{cases}$$

(7.5)

where

$$V(u, v) \triangleq \frac{1}{2\pi} \int_0^u \int_0^{vx/u} \exp\left(-\frac{x^2 + y^2}{2}\right) dy \, dx$$

(7.6)

is a bivariate Gaussian integral that has been tabulated in Ref. 10.

C. Independent Gaussian (÷) Gaussian (Both Have Nonzero Mean)

Let $X_1 \in N_1(\overline{X}_1, \sigma_1^2)$ and $X_2 \in N_1(\overline{X}_2, \sigma_2^2)$ be independent Gaussian RVs. Then, the ratio of these RVs $X = X_2 / X_1$ has the following statistical properties

$$p_X(x) = \frac{\sigma_1 \sigma_2}{\pi(\sigma_1^2 x^2 + \sigma_2^2)} \exp\left[-\frac{1}{2}\left(\frac{\overline{X}_1^2}{\sigma_1^2} + \frac{\overline{X}_2^2}{\sigma_2^2}\right)\right] + \frac{\overline{X}_1 \sigma_2^2 + \overline{X}_2 \sigma_1^2 x}{\sqrt{2\pi}(\sigma_1^2 x^2 + \sigma_2^2)^{3/2}}$$

$$\times \exp\left(-\frac{(\overline{X}_2 - \overline{X}_1 x)^2}{2(\sigma_1^2 x^2 + \sigma_2^2)}\right)\left[1 - 2Q\left(\frac{\overline{X}_1 \sigma_2^2 + \overline{X}_2 \sigma_1^2 x}{\sigma_1 \sigma_2 (\sigma_1^2 x^2 + \sigma_2^2)^{1/2}}\right)\right]$$

(7.7)

$$P_X(x) = \text{------}\qquad(7.8)$$

D. Dependent Gaussian (÷) Gaussian (Both Have Zero Mean)

$$p_X(x) = \frac{\sigma_1\sigma_2\left(1-\rho^2\right)^{1/2}}{\pi\left(\sigma_1^2 x^2 - 2\rho\sigma_1\sigma_2 x + \sigma_2^2\right)}\qquad(7.9)$$

$$P_X(x) = \frac{1}{2} + \frac{1}{\pi}\tan^{-1}\left[\frac{1}{\left(1-\rho^2\right)^{1/2}}\left(\frac{\sigma_1 x}{\sigma_2} - \rho\right)\right]\qquad(7.10)$$

$$\Psi_X(\omega) = \exp\left[\frac{\sigma_2}{\sigma_1}\left(j\omega\rho - \left(1-\rho^2\right)^{1/2}|\omega|\right)\right]\qquad(7.11)$$

E. Dependent Gaussian (÷) Gaussian (One Has Zero Mean)

$$p_X(x) = \frac{\sigma_1\sigma_2\left(1-\rho^2\right)^{1/2}}{\pi\left(\sigma_1^2 x^2 - 2\rho\sigma_1\sigma_2 x + \sigma_2^2\right)}\exp\left(-\frac{\overline{X}_1^2}{2\sigma_1^2\left(1-\rho^2\right)}\right)$$

$$+\frac{\overline{X}_1\sigma_2^2 - \overline{X}_1\rho\sigma_1\sigma_2 x}{\sqrt{2\pi}\left(\sigma_1^2 x^2 - 2\rho\sigma_1\sigma_2 x + \sigma_2^2\right)^{3/2}}$$

$$\times\exp\left(-\frac{\overline{X}_1^2 x^2}{2\left(\sigma_1^2 x^2 - 2\rho\sigma_1\sigma_2 x + \sigma_2^2\right)}\right)$$

$$\times\left[1 - 2Q\left(\frac{\overline{X}_1\sigma_2^2 - \overline{X}_1\rho\sigma_1\sigma_2 x}{\sigma_1\sigma_2\left(1-\rho^2\right)^{1/2}\left(\sigma_1^2 x^2 - 2\rho\sigma_1\sigma_2 x + \sigma_2^2\right)^{1/2}}\right)\right]$$

$$(7.12)$$

$$P_X(x) = \text{-------}\qquad(7.13)$$

F. Dependent Gaussian (+) Gaussian (Both Have Nonzero Mean)

$$p_X(x) = \frac{\sigma_1\sigma_2(1-\rho^2)^{1/2}}{\pi(\sigma_1^2 x^2 - 2\rho\sigma_1\sigma_2 x + \sigma_2^2)} \exp\left[-\frac{1}{2(1-\rho^2)}\left(\frac{\overline{X}_1^2}{\sigma_1^2} - 2\rho\frac{\overline{X}_1}{\sigma_1}\frac{\overline{X}_2}{\sigma_2} + \frac{\overline{X}_2^2}{\sigma_2^2}\right)\right]$$

$$+ \frac{\overline{X}_1\sigma_2^2 - \overline{X}_2\rho\sigma_1\sigma_2 + (\overline{X}_2\sigma_1^2 - \overline{X}_1\rho\sigma_1\sigma_2)x}{\sqrt{2\pi}(\sigma_1^2 x^2 - 2\rho\sigma_1\sigma_2 x + \sigma_2^2)^{3/2}}$$

$$\times \exp\left(-\frac{(\overline{X}_2 - \overline{X}_1 x)^2}{2(\sigma_1^2 x^2 - 2\rho\sigma_1\sigma_2 x + \sigma_2^2)}\right)$$

$$\times \left[1 - 2Q\left(\frac{\overline{X}_1\sigma_2^2 - \overline{X}_2\rho\sigma_1\sigma_2 + (\overline{X}_2\sigma_1^2 - \overline{X}_1\rho\sigma_1\sigma_2)x}{\sigma_1\sigma_2(1-\rho^2)^{1/2}(\sigma_1^2 x^2 - 2\rho\sigma_1\sigma_2 x + \sigma_2^2)^{1/2}}\right)\right]$$

$$\tag{7.14}$$

$$P_X(x) = - - - - - - \tag{7.15}$$

G. Independent Gaussian (Zero Mean) (+) Rayleigh

Let $X \in N_1(0, \sigma_1^2)$ and $R = \|\mathbf{X}^{(2)}\|$ with $\mathbf{X}^{(2)} \in N_n(0, \sigma_2^2)$ be independent Gaussian and Rayleigh RVs. Then, the ratio $Z = X/R$ has the following statistical properties.

1. $n = 1$

$$p_Z(z) = \frac{\sigma_1\sigma_2}{\pi(\sigma_2^2 z^2 + \sigma_1^2)} \tag{7.16}$$

$$P_Z(z) = \frac{1}{2} + \frac{1}{\pi}\tan^{-1}\left(\frac{\sigma_2 z}{\sigma_1}\right) \tag{7.17}$$

$$\Psi_Z(\omega) = \exp\left(-\frac{\sigma_1}{\sigma_2}|\omega|\right) \tag{7.18}$$

2. $n = 2$

$$p_Z(z) = \frac{\sigma_1^2 \sigma_2}{2(\sigma_2^2 z^2 + \sigma_1^2)^{3/2}} \qquad (7.19)$$

$$P_Z(z) = \frac{1}{2} + \frac{1}{2} \frac{\sigma_2 z}{(\sigma_2^2 z^2 + \sigma_1^2)^{1/2}} \qquad (7.20)$$

3. $n = 3$

$$p_Z(z) = \frac{2\sigma_1^3 \sigma_2}{\pi(\sigma_2^2 z^2 + \sigma_1^2)^2} \qquad (7.21)$$

$$P_Z(z) = \frac{1}{2} + \frac{1}{\pi} \tan^{-1}\left(\frac{\sigma_2 z}{\sigma_1}\right) + \frac{\sigma_1 \sigma_2 z}{\pi(\sigma_2^2 z^2 + \sigma_1^2)} \qquad (7.22)$$

$$\Psi_Z(\omega) = \left(1 + \frac{\sigma_1}{\sigma_2}|\omega|\right) \exp\left(-\frac{\sigma_1}{\sigma_2}|\omega|\right) \qquad (7.23)$$

4. $n = 2m$

$$p_Z(z) = \frac{\sigma_1^{2m} \sigma_2 \Gamma(m + 1/2)}{\sqrt{\pi}(m-1)!(\sigma_2^2 z^2 + \sigma_1^2)^{m+1/2}} \qquad (7.24)$$

$$P_Z(z) = \frac{1}{2} + \frac{1}{2} \frac{\sigma_2 z}{(\sigma_2^2 z^2 + \sigma_1^2)^{1/2}} \sum_{i=1}^{m} \frac{1}{4^{i-1}} \binom{2i-2}{i-1} \left(\frac{\sigma_1^2}{\sigma_2^2 z^2 + \sigma_1^2}\right)^{i-1} \qquad (7.25)$$

5. $n = 2m + 1$

$$p_Z(z) = \frac{\sigma_1^{2m+1} \sigma_2 m!}{\sqrt{\pi}\Gamma(m + 1/2)(\sigma_2^2 z^2 + \sigma_1^2)^{m+1}} \qquad (7.26)$$

$$P_Z(z) = \frac{1}{2} + \frac{1}{\pi} \tan^{-1}\left(\frac{\sigma_2 z}{\sigma_1}\right) + \frac{1}{\pi}\left(\frac{\sigma_2 z}{\sigma_1}\right) \sum_{i=1}^{m} \frac{4^{i-1}}{(2i-1)\binom{2i-2}{i-1}} \left(\frac{\sigma_1^2}{\sigma_2^2 z^2 + \sigma_1^2}\right)^{i} \qquad (7.27)$$

H. Independent Gaussian (Zero Mean) (+) Rice

Let $X \in N_1(0, \sigma_1^2)$ and $R = \|\mathbf{X}^{(2)}\|$ with $\mathbf{X}^{(2)} \in N_n(\overline{\mathbf{X}}^{(2)}, \sigma_2^2)$ be independent Gaussian and Rice RVs. Then, the ratio $Z = X / R$ has the following statistical properties.

1. $n = 1$

$$p_Z(z) = \frac{\sigma_1 \sigma_2}{\pi(\sigma_2^2 z^2 + \sigma_1^2)} \exp\left(-\frac{a_2^2}{2\sigma_2^2}\right) {}_1F_1\left(1, \frac{1}{2}; \frac{a_2^2 \sigma_1^2}{2\sigma_2^2(\sigma_2^2 z^2 + \sigma_1^2)}\right) \quad (7.28)$$

$$P_Z(z) = \begin{cases} \dfrac{1}{\pi} \tan^{-1}\left(\dfrac{\sigma_1}{\sigma_2|z|}\right) - 2V\left(\dfrac{a_2|z|}{\sqrt{\sigma_2^2 z^2 + \sigma_1^2}}, \dfrac{a_2(\sigma_1 / \sigma_2)}{\sqrt{\sigma_2^2 z^2 + \sigma_1^2}}\right), z < 0 \\[4mm] 1 - \dfrac{1}{\pi} \tan^{-1}\left(\dfrac{\sigma_1}{\sigma_2 z}\right) + 2V\left(\dfrac{a_2 z}{\sqrt{\sigma_2^2 z^2 + \sigma_1^2}}, \dfrac{a_2(\sigma_1 / \sigma_2)}{\sqrt{\sigma_2^2 z^2 + \sigma_1^2}}\right), z > 0 \end{cases} \quad (7.29)$$

2. $n = 2$

$$p_Z(z) = \frac{\sigma_1^2 \sigma_2}{2(\sigma_2^2 z^2 + \sigma_1^2)^{3/2}} \exp\left[-\frac{a_2^2}{4\sigma_2^2}\left(\frac{2\sigma_2^2 z^2 + \sigma_1^2}{\sigma_2^2 z^2 + \sigma_1^2}\right)\right]$$

$$\times\left[\left(1 + \frac{a_2^2 \sigma_1^2}{2\sigma_2^2(\sigma_2^2 z^2 + \sigma_1^2)}\right) I_0\left(\frac{a_2^2 \sigma_1^2}{4\sigma_2^2(\sigma_2^2 z^2 + \sigma_1^2)}\right)\right.$$

$$\left. + \frac{a_2^2 \sigma_1^2}{2\sigma_2^2(\sigma_2^2 z^2 + \sigma_1^2)} I_1\left(\frac{a_2^2 \sigma_1^2}{4\sigma_2^2(\sigma_2^2 z^2 + \sigma_1^2)}\right)\right] \quad (7.30)$$

$$P_Z(z) = \begin{cases} Q_1(\alpha, \beta) - \left(\dfrac{\sigma_2 \beta}{a_2}\right) \exp\left(-\dfrac{\alpha^2 + \beta^2}{2}\right) I_0(\alpha\beta), z < 0 \\[4mm] 1 - Q_1(\alpha, \beta) + \left(\dfrac{\sigma_2 \beta}{a_2}\right) \exp\left(-\dfrac{\alpha^2 + \beta^2}{2}\right) I_0(\alpha\beta), z \geq 0 \end{cases} \quad (7.31)$$

where

$$\alpha \triangleq \frac{a_2}{2\sigma_2}\left[1-\left(\frac{\sigma_2^2 z^2}{\sigma_2^2 z^2 + \sigma_1^2}\right)^{1/2}\right], \quad \beta \triangleq \frac{a_2}{2\sigma_2}\left[1+\left(\frac{\sigma_2^2 z^2}{\sigma_2^2 z^2 + \sigma_1^2}\right)^{1/2}\right] \quad (7.32)$$

3. $n = 2m$

$$p_Z(z) = \frac{\sigma_1^{2m}\sigma_2 \Gamma(m+1/2)}{\sqrt{\pi}(m-1)!\left(\sigma_2^2 z^2 + \sigma_1^2\right)^{m+1/2}} \exp\left(-\frac{a_2^2}{2\sigma_2^2}\right) {}_1F_1\left(m+\frac{1}{2}, m; \frac{a_2^2 \sigma_1^2}{2\sigma_2^2\left(\sigma_2^2 z^2 + \sigma_1^2\right)}\right)$$

$$(7.33)$$

$$P_Z(z) = \begin{cases} Q_1(\alpha,\beta) - \left(\dfrac{\sigma_2\beta}{a_2}\right)\exp\left(-\dfrac{\alpha^2+\beta^2}{2}\right)I_0(\alpha\beta) - \dfrac{1}{2}\left(\dfrac{\sigma_2^2 z^2}{\sigma_2^2 z^2 + \sigma_1^2}\right)^{1/2} \\ \quad \times \displaystyle\sum_{i=2}^{m}\dfrac{1}{4^{i-1}}\binom{2i-2}{i-1}\left(\dfrac{1}{z^2+1}\right)^{i-1} {}_1F_1\left(\dfrac{1}{2},i;-\dfrac{a_2^2(\sigma_1/\sigma_2)^2}{2(\sigma_2^2 z^2 + \sigma_1^2)}\right), \quad z<0 \\[4mm] 1 - Q_1(\alpha,\beta) + \left(\dfrac{\sigma_2\beta}{a_2}\right)\exp\left(-\dfrac{\alpha^2+\beta^2}{2}\right)I_0(\alpha\beta) + \dfrac{1}{2}\left(\dfrac{\sigma_2^2 z^2}{\sigma_2^2 z^2 + \sigma_1^2}\right)^{1/2} \\ \quad \times \displaystyle\sum_{i=2}^{m}\dfrac{1}{4^{i-1}}\binom{2i-2}{i-1}\left(\dfrac{1}{z^2+1}\right)^{i-1} {}_1F_1\left(\dfrac{1}{2},i;-\dfrac{a_2^2(\sigma_1/\sigma_2)^2}{2(\sigma_2^2 z^2 + \sigma_1^2)}\right), \quad z\geq 0 \end{cases}$$

$$(7.34)$$

4. $n = 2m+1$

$$p_Z(z) = \frac{\sigma_1^{2m+1}\sigma_2 m!}{\sqrt{\pi}\,\Gamma(m+1/2)\left(\sigma_2^2 z^2 + \sigma_1^2\right)^{m+1}} \exp\left(-\frac{a_2^2}{2\sigma_2^2}\right)$$

$$\times {}_1F_1\left(m+1, m+\frac{1}{2}; \frac{a_2^2 \sigma_1^2}{2\sigma_2^2\left(\sigma_2^2 z^2 + \sigma_1^2\right)}\right)$$

$$(7.35)$$

$$P_Z(z) = ------ \quad (7.36)$$

I. Independent Rayleigh (+) Rayleigh

Let $R_1 = \left\|\mathbf{X}^{(1)}\right\|$ and $R_2 = \left\|\mathbf{X}^{(2)}\right\|$ be independent Rayleigh RVs corresponding to Gaussian vectors $\mathbf{X}^{(1)} \in N_{n_1}(\mathbf{0}, \sigma_1^2)$ and $\mathbf{X}^{(2)} \in N_{n_2}(\mathbf{0}, \sigma_2^2)$. Then, the ratio $R = R_2 / R_1$ has the following statistical properties.

1. $n_1 = n_2 = 1$

$$p_R(r) = \frac{2\sigma_1\sigma_2}{\pi(\sigma_1^2 r^2 + \sigma_2^2)}, r \geq 0 \tag{7.37}$$

$$P_R(r) = \frac{2}{\pi} \tan^{-1}\left(\frac{\sigma_1 r}{\sigma_2}\right), r \geq 0 \tag{7.38}$$

$$\Psi_R(\omega) = \exp\left(-\frac{\sigma_2}{\sigma_1}|\omega|\right) \tag{7.39}$$

2. $n_1 = 1, n_2 = 2$

$$p_R(r) = \frac{\sigma_1^2 \sigma_2 r}{(\sigma_1^2 r^2 + \sigma_2^2)^{3/2}}, r \geq 0 \tag{7.40}$$

$$P_R(r) = 1 - \left(\frac{\sigma_2^2}{\sigma_1^2 r^2 + \sigma_2^2}\right)^{1/2}, r \geq 0 \tag{7.41}$$

3. $n_1 = 1, n_2 = 2m$

$$p_R(r) = \frac{2\Gamma(m+1/2)\sigma_1^{2m}\sigma_2 r^{2m-1}}{\sqrt{\pi}(m-1)!(\sigma_1^2 r^2 + \sigma_2^2)^{m+1/2}}, r \geq 0 \tag{7.42}$$

$$P_R(r) = 1 - \left(\frac{\sigma_2^2}{\sigma_1^2 r^2 + \sigma_2^2}\right)^{1/2} \sum_{i=1}^{m} \frac{1}{4^{i-1}}\binom{2i-2}{i-1}\left(\frac{\sigma_1^2 r^2}{\sigma_1^2 r^2 + \sigma_2^2}\right)^{i-1}, r \geq 0 \tag{7.43}$$

4. $n_1 = n_2 = 2$

$$p_R(r) = \frac{2\sigma_1^2 \sigma_2^2 r}{(\sigma_1^2 r^2 + \sigma_2^2)^2}, r \geq 0 \tag{7.44}$$

$$P_R(r) = 1 - \frac{\sigma_2^2}{\sigma_1^2 r^2 + \sigma_2^2}, r \ge 0 \tag{7.45}$$

$$E\{R\} = \frac{\pi \sigma_2}{2\sigma_1} \tag{7.46}$$

5. $n_1 = n_2 = n$

$$p_R(r) = \frac{2(n-1)! \sigma_1^n \sigma_2^n r^{n-1}}{[\Gamma(n/2)]^2 (\sigma_1^2 r^2 + \sigma_2^2)^n}, r \ge 0 \tag{7.47}$$

$$P_R(r)\big|_n = P_R(r)\big|_{n-2} - \frac{(\sigma_2^2 - \sigma_1^2 r^2) \Gamma\left(\dfrac{n-1}{2}\right)(2\sigma_1 \sigma_2 r)^{n-2}}{\sqrt{\pi}(n-2)\Gamma\left(\dfrac{n-2}{2}\right)(\sigma_1^2 r^2 + \sigma_2^2)^{n-1}}, r \ge 0$$

(recursive form)
$$\tag{7.48}$$

$$E\{R^k\} = \left(\frac{\sigma_1}{\sigma_2}\right)^k \frac{\Gamma\left(\dfrac{n+k}{2}\right)\Gamma\left(\dfrac{n-k}{2}\right)}{[\Gamma(n/2)]^{1/2}}, 0 \le k \le n \tag{7.49}$$

6. $n_1 = 2m_1, n_2 = 2m_2$

$$p_R(r) = \frac{2(m_1 + m_2 - 1)! \sigma_1^{2m_2} \sigma_2^{2m_1} r^{2m_2-1}}{(m_1 - 1)!(m_2 - 1)!(\sigma_1^2 r^2 + \sigma_2^2)^{m_1+m_2}}, r \ge 0 \tag{7.50}$$

$$P_R(r) = 1 - \frac{(m_1 + m_2 - 1)!}{(m_1 - 1)!(m_2 - 1)!} \sum_{i=0}^{m_2-1} \binom{m_2 - 1}{i} \frac{(-1)^i}{m_1 + i} \left(\frac{\sigma_2^2}{\sigma_1^2 r^2 + \sigma_2^2}\right)^{m_1+i}, r \ge 0 \tag{7.51}$$

7. $n_1 = 2m_1 + 1, n_2 = 2m_2 + 1$

$$p_R(r) = \frac{2(m_1 + m_2)! \sigma_1^{2m_2+1} \sigma_2^{2m_1+1} r^{2m_2}}{\Gamma(m_1 + 1/2)\Gamma(m_2 + 1/2)(\sigma_1^2 r^2 + \sigma_2^2)^{m_1+m_2+1}}, r \ge 0 \tag{7.52}$$

$$P_R(r) = - - - - - - \tag{7.53}$$

8. $n_1 = 2m_1 + 1, n_2 = 2m_2$

$$p_R(r) = \frac{2\Gamma(m_1 + m_2 + 1/2)\sigma_1^{2m_2}\sigma_2^{2m_1+1}r^{2m_2-1}}{\Gamma(m_1 + 1/2)(m_2 - 1)!(\sigma_1^2 r^2 + \sigma_2^2)^{m_1+m_2+1/2}}, r \geq 0 \qquad (7.54)$$

$$P_R(r) = 1 - \left(\frac{\sigma_2^2}{\sigma_1^2 r^2 + \sigma_2^2}\right)^{1/2} - \left(\frac{\sigma_2^2}{\sigma_1^2 r^2 + \sigma_2^2}\right)^{m_1+1/2} \left\{ \sum_{i=0}^{m_2-2} \left(\frac{\sigma_1^2 r^2}{\sigma_1^2 r^2 + \sigma_2^2}\right)^{m_2-i-1} \right.$$

$$\left. \times \frac{\Gamma(m_1 + m_2 - i - 1/2)}{\Gamma(m_1 + 1/2)(m_2 - i - 1)!} - \frac{\sigma_1^2 r^2}{\sigma_2^2} \sum_{i=0}^{m_1-1} \left(\frac{\sigma_1^2 r^2 + \sigma_2^2}{\sigma_2^2}\right)^i \right\}, r \geq 0 \qquad (7.55)$$

J. Dependent Rayleigh (+) Rayleigh

Let $R_1 = \|\mathbf{X}^{(1)}\|$ and $R_2 = \|\mathbf{X}^{(2)}\|$ be independent Rayleigh RVs corresponding to Gaussian vectors $\mathbf{X}^{(1)} \in N_n(\overline{\mathbf{X}}^{(1)}, \sigma_1^2)$ and $\mathbf{X}^{(2)} \in N_n(\overline{\mathbf{X}}^{(2)}, \sigma_2^2)$. Then, the ratio $R = R_2 / R_1$ has the following statistical properties.

1. $n = 1$

$$p_R(r) = \frac{2\sigma_1\sigma_2(1 - \rho^2)(\sigma_1^2 r^2 + \sigma_2^2)}{\pi\left[(\sigma_1^2 r^2 + \sigma_2^2)^2 - 4\rho^2\sigma_1^2\sigma_2^2 r^2\right]}, r \geq 0 \qquad (7.56)$$

$$P_R(r) = \frac{1}{2} - \frac{1}{\pi}\tan^{-1}\left(\frac{\sigma_2^2 - \sigma_1^2 r^2}{2\sigma_1\sigma_2 r\sqrt{1 - \rho^2}}\right), r \geq 0 \qquad (7.57)$$

2. $n = 2$

$$p_R(r) = \frac{2\left[\sigma_1\sigma_2(1 - \rho^2)\right]^2 r(\sigma_1^2 r^2 + \sigma_2^2)}{\left[(\sigma_1^2 r^2 + \sigma_2^2)^2 - 4\rho^2\sigma_1^2\sigma_2^2 r^2\right]^{3/2}}, r \geq 0 \qquad (7.58)$$

$$P_R(r) = \frac{1}{2} - \frac{1}{2}\frac{\sigma_2^2 - \sigma_1^2 r^2}{\left[4(1 - \rho^2)\sigma_1^2\sigma_2^2 r^2 + (\sigma_2^2 - \sigma_1^2 r^2)^2\right]^{1/2}}, r \geq 0 \qquad (7.59)$$

$$E\{R\} = \frac{\pi}{2}\left(\frac{\sigma_1}{\sigma_2}\right)\left(1-\rho^2\right)^2 {}_2F_1\left(\frac{3}{2},\frac{1}{2};1;\rho^2\right) \tag{7.60}$$

3. $n = 2m$

$$p_R(r) = \frac{2(2m-1)!\left[\sigma_1\sigma_2\left(1-\rho^2\right)\right]^{2m} r^{2m-1}\left(\sigma_1^2 r^2 + \sigma_2^2\right)}{[(m-1)!]^2\left[\left(\sigma_1^2 r^2 + \sigma_2^2\right)^2 - 4\rho^2\sigma_1^2\sigma_2^2 r^2\right]^{m+1/2}}, \quad r \geq 0 \tag{7.61}$$

$$P_R(r)\big|_{2m} = P_R(r)\big|_{2m-2} - \frac{\left(1-\rho^2\right)^{m-1}\left(\sigma_2^2 - \sigma_1^2 r^2\right)\Gamma\left(m-\frac{1}{2}\right)\left(2\sigma_1\sigma_2 r\right)^{2m-2}}{\sqrt{\pi}(2m-2)(m-2)!\left[4\left(1-\rho^2\right)\sigma_1^2\sigma_2^2 r^2 + \left(\sigma_2^2 - \sigma_1^2 r^2\right)^2\right]^{m-1/2}},$$

$$r \geq 0 \quad \text{(recursive form)}$$

$$\tag{7.62}$$

$$E\{R^k\} = \frac{\Gamma\left(m+\dfrac{k}{2}\right)\Gamma\left(m-\dfrac{k}{2}\right)}{[(m-1)!]^2}\left(\frac{\sigma_1}{\sigma_2}\right)^k\left(1-\rho^2\right)^{2m} {}_2F_1\left(m+\frac{k}{2},m-\frac{k}{2};m;\rho^2\right), \tag{7.63}$$

$$0 \leq k < 2m$$

4. $n = 2m+1$

$$p_R(r) = \frac{2(2m)!\left[\sigma_1\sigma_2\left(1-\rho^2\right)\right]^{2m+1} r^{2m}\left(\sigma_1^2 r^2 + \sigma_2^2\right)}{\left[\Gamma\left(m+\dfrac{1}{2}\right)\right]^2\left[\left(\sigma_1^2 r^2 + \sigma_2^2\right)^2 - 4\rho^2\sigma_1^2\sigma_2^2 r^2\right]^{m+1}}, \quad r \geq 0 \tag{7.64}$$

$$P_R(r)\big|_{2m+1} = P_R(r)\big|_{2m-1}$$

$$- \frac{\left(1-\rho^2\right)^{m-1/2}\left(\sigma_2^2 - \sigma_1^2 r^2\right)(m-1)!\left(2\sigma_1\sigma_2 r\right)^{2m-1}}{\sqrt{\pi}(2m-1)\Gamma(m-1/2)\left[4\left(1-\rho^2\right)\sigma_1^2\sigma_2^2 r^2 + \left(\sigma_2^2 - \sigma_1^2 r^2\right)^2\right]^{m}}, \tag{7.65}$$

$$r \geq 0 \quad \text{(recursive form)}$$

$$E\{R^k\} = \frac{\Gamma\left(m+\dfrac{k+1}{2}\right)\Gamma\left(m-\dfrac{k-1}{2}\right)}{[\Gamma(m+1/2)]^2}\left(\frac{\sigma_1}{\sigma_2}\right)^k (1-\rho^2)^{2m+1}$$

$$\times\,_2F_1\left(m+\frac{k+1}{2}, m-\frac{k-1}{2}; m+\frac{1}{2};\rho^2\right), 0\le k < 2m+1 \tag{7.66}$$

K. Independent Rice (+) Rayleigh

Let $R_1 = \|\mathbf{X}^{(1)}\|$ and $R_2 = \|\mathbf{X}^{(2)}\|$ be independent Rice and Rayleigh RVs corresponding to Gaussian vectors $\mathbf{X}^{(1)} \in N_{n_1}(0,\sigma_1^2)$ and $\mathbf{X}^{(2)} \in N_{n_2}(\overline{\mathbf{X}}^{(2)},\sigma_2^2)$. Then, the ratio $R = R_2/R_1$ has the following statistical properties.

1. $n_1 = n_2 = 1$

$$p_R(r) = \frac{2\sigma_1\sigma_2}{\pi\left(\sigma_1^2 r^2 + \sigma_2^2\right)}\exp\left(-\frac{a_2^2}{2\sigma_2^2}\right)\,_1F_1\left(1;\frac{1}{2};\frac{a_2^2\sigma_1^2 r^2}{2\sigma_2^2\left(\sigma_1^2 r^2+\sigma_2^2\right)}\right), r\ge 0 \tag{7.67}$$

$$p_R(r) = \frac{2}{\pi}\tan^{-1}\left(\frac{\sigma_1 r}{\sigma_2}\right) - 4V\left(\frac{a_2}{\sqrt{\sigma_1^2 r^2 + \sigma_2^2}}, \frac{a_2(\sigma_2/\sigma_1)r}{\sqrt{\sigma_1^2 r^2 + \sigma_2^2}}\right), r\ge 0 \tag{7.68}$$

2. $n_1 = 1, n_2 = 2$

$$p_R(r) = \frac{\sigma_1^2\sigma_2 r}{\left(\sigma_1^2 r^2 + \sigma_2^2\right)^{3/2}}\exp\left(-\frac{\alpha^2+\beta^2}{2}\right)\left[(1+2\alpha\beta)I_0(\alpha\beta) + 2\alpha\beta I_1(\alpha\beta)\right], r\ge 0 \tag{7.69}$$

$$p_R(r) = 2\left[Q(\alpha,\beta) - \left(\frac{\sigma_2\beta}{a_2}\right)\exp\left(-\frac{\alpha^2+\beta^2}{2}\right)I_0(\alpha\beta)\right], r\ge 0 \tag{7.70}$$

where

$$\alpha \triangleq \frac{a_2}{2\sigma_2}\left[1-\left(\frac{\sigma_2^2}{\sigma_1^2 r^2 + \sigma_2^2}\right)^{1/2}\right], \quad \beta \triangleq \frac{a_2}{2\sigma_2}\left[1+\left(\frac{\sigma_2^2}{\sigma_1^2 r^2 + \sigma_2^2}\right)^{1/2}\right] \tag{7.71}$$

3. $n_1 = 1, n_2 = 2m$

$$p_R(r) = \frac{2\Gamma(m+1/2)\sigma_1^{2m}\sigma_2 r^{2m-1}}{\sqrt{\pi}(m-1)!(\sigma_1^2 r^2 + \sigma_2^2)^{m+1/2}} \exp\left(-\frac{a_2^2}{2\sigma_2^2}\right) {}_1F_1\left(m+\frac{1}{2};m;\frac{a_2^2\sigma_1^2 r^2}{2\sigma_2^2(\sigma_1^2 r^2 + \sigma_2^2)}\right)$$

$$,r \geq 0$$
$$(7.72)$$

$$P_R(r) = 2\left[Q(\alpha,\beta) - \left(\frac{\sigma_2\beta}{a_2}\right)\exp\left(-\frac{\alpha^2+\beta^2}{2}\right)I_0(\alpha\beta)\right] - \left(\frac{\sigma_2^2}{\sigma_1^2 r^2 + \sigma_2^2}\right)^{1/2}$$

$$\times \exp\left[-\frac{a_2^2}{2(\sigma_1^2 r^2 + \sigma_2^2)}\right]\sum_{i=2}^{m}\frac{1}{4^{i-1}}\binom{2i-2}{i-1}\left(\frac{\sigma_1^2 r^2}{\sigma_1^2 r^2 + \sigma_2^2}\right)^{i-1}$$

$$\times {}_1F_1\left(\frac{1}{2};i;-\frac{a_2^2\sigma_1^2 r^2}{2\sigma_2^2(\sigma_1^2 r^2 + \sigma_2^2)}\right), r \geq 0$$

$$(7.73)$$

4. $n_1 = n_2 = 2$

$$p_R(r) = \frac{2\sigma_1^2\sigma_2^2 r}{(\sigma_1^2 r^2 + \sigma_2^2)^2}\exp\left[-\frac{a_2^2}{2(\sigma_1^2 r^2 + \sigma_2^2)}\right]\left[1 + \frac{a_2^2\sigma_1^2 r^2}{2\sigma_2^2(\sigma_1^2 r^2 + \sigma_2^2)}\right], r \geq 0$$

$$(7.74)$$

$$P_R(r) = \frac{\sigma_1^2 r^2}{\sigma_1^2 r^2 + \sigma_2^2}\exp\left[-\frac{a_2^2}{2(\sigma_1^2 r^2 + \sigma_2^2)}\right], r \geq 0 \qquad (7.75)$$

5. $n_1 = 2m_1, n_2 = 2m_2$

$$p_R(r) = \frac{2(m_1 + m_2 - 1)!\sigma_1^{2m_2}\sigma_2^{2m_1} r^{2m_2-1}}{(m_1-1)!(m_2-1)!(\sigma_1^2 r^2 + \sigma_2^2)^{m_1+m_2}}\exp\left(-\frac{a_2^2}{2\sigma_2^2}\right)$$

$$\times {}_1F_1\left(m_1 + m_2;m_2;\frac{a_2^2\sigma_1^2 r^2}{2\sigma_2^2(\sigma_1^2 r^2 + \sigma_2^2)}\right), r \geq 0$$

$$(7.76)$$

$$P_R(r) = \left(\frac{\sigma_1^2 r^2}{\sigma_1^2 r^2 + \sigma_2^2}\right)^{m_2} \exp\left[-\frac{a_2^2}{2(\sigma_1^2 r^2 + \sigma_2^2)}\right] \sum_{i=0}^{m_1-1} \sum_{l=i}^{m_1-1} \frac{1}{i! \, 2^i}\binom{m_2+l-1}{l-i}$$

$$\times \left(\frac{a_2^2 \sigma_1^2 r^2}{\sigma_2^2(\sigma_1^2 r^2 + \sigma_2^2)}\right)^i \left(\frac{\sigma_2^2}{\sigma_1^2 r^2 + \sigma_2^2}\right)^l, \, r \geq 0 \tag{7.77}$$

6. $n_1 = 2m_1 + 1, n_2 = 2m_2$

$$P_R(r) = \frac{2\Gamma\left(m_1 + m_2 + \frac{1}{2}\right)\sigma_1^{2m_2}\sigma_2^{2m_1+1}r^{2m_2-1}}{\Gamma(m_1 + 1/2)(m_2 - 1)!(\sigma_1^2 r^2 + \sigma_2^2)^{m_1+m_2+1/2}} \exp\left(-\frac{a_2^2}{2\sigma_2^2}\right)$$

$$\times \, _1F_1\left(m_1 + m_2 + \frac{1}{2}; m_2; \frac{a_2^2 \sigma_1^2 r^2}{2\sigma_2^2(\sigma_1^2 r^2 + \sigma_2^2)}\right), \, r \geq 0 \tag{7.78}$$

$$P_R(r) = 2\left[Q(\alpha, \beta) - \left(\frac{\sigma_2 \beta}{a_2}\right)\exp\left(-\frac{\alpha^2 + \beta^2}{2}\right)I_0(\alpha\beta)\right] - \left(\frac{\sigma_2^2}{\sigma_1^2 r^2 + \sigma_2^2}\right)^{m_1+1/2}$$

$$\times \exp\left(-\frac{a_2^2}{2\sigma_2^2}\right)\left\{\sum_{i=0}^{m_2-2}\left(\frac{\sigma_1^2 r^2}{\sigma_1^2 r^2 + \sigma_2^2}\right)^{m_2-i-1} \frac{\Gamma\left(m_1 + m_2 - i - \frac{1}{2}\right)}{(m_2 - i - 1)! \, \Gamma(m_1 + 1/2)}\right.$$

$$\times \, _1F_1\left(m_1 + m_2 - i - \frac{1}{2}; m_2 - i; \frac{a_2^2 \sigma_1^2 r^2}{2\sigma_2^2(\sigma_1^2 r^2 + \sigma_2^2)}\right) - \left(\frac{\sigma_1 r}{\sigma_2}\right)^{2m_1-2}$$

$$\times \sum_{i=0}^{m_1-1}\left(\frac{\sigma_1^2 r^2 + \sigma_2^2}{\sigma_2^2}\right)^i \, _1F_1\left(m_1 - i + \frac{1}{2}; 1; \frac{a_2^2 \sigma_1^2 r^2}{2\sigma_2^2(\sigma_1^2 r^2 + \sigma_2^2)}\right)\right\}, \, r \geq 0 \tag{7.79}$$

L. Independent Rice (÷) Rice

Let $R_1 = \|\mathbf{X}^{(1)}\|$ and $R_2 = \|\mathbf{X}^{(2)}\|$ be independent Rice RVs corresponding to Gaussian vectors $\mathbf{X}^{(1)} \in N_{n_1}\left(\overline{\mathbf{X}}^{(1)}, \sigma_1^2\right)$ and $\mathbf{X}^{(2)} \in N_{n_2}\left(\overline{\mathbf{X}}^{(2)}, \sigma_2^2\right)$. Then, the ratio $R = R_2 / R_1$ has the following statistical properties.

1. $n_1 = n_2 = 1$

$$p_R(r) = \frac{2\sigma_1\sigma_2}{\sigma_1^2 r^2 + \sigma_2^2} \exp\left[-\frac{1}{2}\left(\frac{a_1^2}{2\sigma_1^2} + \frac{a_2^2}{2\sigma_2^2}\right)\right] \sum_{i=0}^{\infty} \sum_{l=0}^{\infty} \binom{i+l}{l} \frac{1}{\Gamma(i+1/2)\Gamma(l+1/2)}$$

$$\times \left(\frac{a_1^2\sigma_2^2}{2\sigma_1^2}\right)^i \left(\frac{a_2^2\sigma_1^2 r^2}{2\sigma_2^2}\right)^l \frac{1}{\left(\sigma_1^2 r^2 + \sigma_2^2\right)^{i+l}}, r \geq 0$$

(7.80)

$$P_R(r) = \frac{2}{\pi} \tan^{-1}\left(\frac{\sigma_1 r}{\sigma_2}\right) + 2V\left(\frac{a_1 r + a_2}{\sqrt{\sigma_1^2 r^2 + \sigma_2^2}}, \frac{a_1(\sigma_2/\sigma_1) - a_2(\sigma_1/\sigma_2)r}{\sqrt{\sigma_1^2 r^2 + \sigma_2^2}}\right)$$

$$+ 2V\left(\frac{a_1 r - a_2}{\sqrt{\sigma_1^2 r^2 + \sigma_2^2}}, \frac{a_1(\sigma_2/\sigma_1) + a_2(\sigma_1/\sigma_2)r}{\sqrt{\sigma_1^2 r^2 + \sigma_2^2}}\right), r \geq 0$$

(7.81)

2. $n_1 = n_2 = 2$

$$p_R(r) = \frac{2\sigma_1^2\sigma_2^2 r}{\left(\sigma_1^2 r^2 + \sigma_2^2\right)^2} \exp\left[-\frac{a_1^2 r^2 + a_2^2}{2(\sigma_1^2 r^2 + \sigma_2^2)}\right]\left[\left(1 + \frac{a_1^2\sigma_1^4 + a_2^2\sigma_2^4 r^2}{2\sigma_1^2\sigma_2^2(\sigma_1^2 r^2 + \sigma_2^2)}\right)\right.$$

$$\left. \times I_0\left(\frac{a_1 a_2 r}{\sigma_1^2 r^2 + \sigma_2^2}\right) + \left(\frac{a_1 a_2 r}{\sigma_1^2 r^2 + \sigma_2^2}\right)I_1\left(\frac{a_1 a_2 r}{\sigma_1^2 r^2 + \sigma_2^2}\right)\right], r \geq 0$$

(7.82)

$$P_R(r) = Q(\alpha, \beta) - \left(\frac{\sigma_2^2\alpha^2}{a_1^2 r^2}\right)\exp\left(-\frac{\alpha^2 + \beta^2}{2}\right)I_0(\alpha\beta), r \geq 0$$

(7.83)

where

$$\alpha \triangleq \left(\frac{a_1^2 r^2}{\sigma_1^2 r^2 + \sigma_2^2}\right)^{1/2}, \quad \beta \triangleq \left(\frac{a_2^2}{\sigma_1^2 r^2 + \sigma_2^2}\right)^{1/2}$$

(7.84)

3. $n_1 = n_2 = 2m$

$$p_R(r) = (-1)^m \frac{a_1 a_2}{\sigma_1^2\sigma_2^2}\left(\frac{r}{a_1 a_2}\right)^m \exp\left[-\frac{1}{2}\left(\frac{a_1^2}{\sigma_1^2} + \frac{a_2^2}{\sigma_2^2}\right)\right]$$

$$\times \frac{d^m}{dc^m}\left\{\frac{1}{2c}\exp\left[\frac{1}{4c}\left(\frac{a_1^2}{\sigma_1^2} + \frac{a_2^2 r^2}{\sigma_2^2}\right)\right]I_{m-1}\left(\frac{a_1 a_2 r}{2c\sigma_1^2\sigma_2^2}\right)\right\}, r \geq 0$$

(7.85)

where

$$c = \frac{\sigma_1^2 r^2 + \sigma_2^2}{2\sigma_1^2 \sigma_2^2} \tag{7.86}$$

$$P_R(r) = Q(\alpha, \beta) - \left(\frac{\sigma_2^2 \alpha^2}{a_1^2 r^2}\right) \exp\left(-\frac{\alpha^2 + \beta^2}{2}\right) I_0(\alpha\beta) + \exp\left[-\frac{a_1^2 r^2 + a_2^2}{2(\sigma_1^2 r^2 + \sigma_2^2)}\right] \tag{7.87}$$

$$\times \sum_{l=1-m}^{m-1} C_l(m-1, m-1; r)\left(\frac{a_2 \sigma_1^2 r}{a_1 \sigma_2^2}\right)^l I_l\left(\frac{a_1 a_2 r}{\sigma_1^2 r^2 + \sigma_2^2}\right), \quad r \geq 0$$

where α and β are defined in (7.84) and

$$C_l(m-1, m-1; r) = \begin{cases} \sum_{i=l}^{m-1} \binom{m-1+i}{i-l}\left(\frac{\sigma_1^2 r^2}{\sigma_1^2 r^2 + \sigma_2^2}\right)^m \left(\frac{\sigma_2^2}{\sigma_1^2 r^2 + \sigma_2^2}\right)^i \\ -\delta_{l0}\left(\frac{\sigma_1^2 r^2}{\sigma_1^2 r^2 + \sigma_2^2}\right), \quad l \geq 0 \\ -C_{-l}\left(m-1, m-1; \frac{\sigma_2}{\sigma_1 r}\right), \quad l < 0 \end{cases} \tag{7.88}$$

with δ_{l0} denoting the Kronecker delta; $\delta_{l0} = 1$ for $l = 0$, $\delta_{l0} = 0$ otherwise.

4. $n_1 = 2m_1, n_2 = 2m_2$

$$P_R(r) = \frac{2\sigma_1^{2m_2}\sigma_2^{2m_1}r^{2m_2-1}}{(\sigma_1^2 r^2 + \sigma_2^2)^{m_1+m_2}} \exp\left[-\frac{1}{2}\left(\frac{a_1^2}{\sigma_1^2} + \frac{a_2^2}{\sigma_2^2}\right)\right] \sum_{i=0}^{\infty} \sum_{l=0}^{\infty} \frac{(m_1+m_2+i+l-1)!}{i!l!(m_1+i-1)!(m_2+l-1)!}$$

$$\times \left(\frac{a_1^2 \sigma_2^2}{2\sigma_1^2}\right)^i \left(\frac{a_2^2 \sigma_1^2 r^2}{2\sigma_2^2}\right)^l \frac{1}{(\sigma_1^2 r^2 + \sigma_2^2)^{i+l}}, \quad r \geq 0$$

$$\tag{7.89}$$

$$P_R(r) = Q(\alpha, \beta) - \left(\frac{\sigma_2^2 \alpha^2}{a_1^2 r^2}\right) \exp\left(-\frac{\alpha^2 + \beta^2}{2}\right) I_0(\alpha\beta) + \exp\left[-\frac{a_1^2 r^2 + a_2^2}{2(\sigma_1^2 r^2 + \sigma_2^2)}\right] \tag{7.90}$$

$$\times \sum_{l=1-m_2}^{m_1-1} C_l(m_1-1, m_2-1; r)\left(\frac{a_2 \sigma_1^2 r}{a_1 \sigma_2^2}\right)^l I_l\left(\frac{a_1 a_2 r}{\sigma_1^2 r^2 + \sigma_2^2}\right), \quad r \geq 0$$

where

$$
C_i(m_1-1,m_2-1;r) = \begin{cases} \sum_{i=l}^{m_1-1}\binom{m_2-1+i}{i-l}\left(\dfrac{\sigma_1^2r^2}{\sigma_1^2r^2+\sigma_2^2}\right)^{m_1}\left(\dfrac{\sigma_2^2}{\sigma_1^2r^2+\sigma_2^2}\right)^i \\[12pt] -\delta_{l0}\left(\dfrac{\sigma_1^2r^2}{\sigma_1^2r^2+\sigma_2^2}\right),\ l\geq 0 \\[12pt] -C_{-l}\left(m_2-1,m_1-1;\dfrac{\sigma_2}{\sigma_1 r}\right),\ l<0 \end{cases}
\tag{7.91}
$$

M. Dependent Rayleigh Ratios

Consider the ratio RV $U_1 = R_3 / R_1$ where $R_1 = \|\mathbf{X}^{(1)}\|$ and $R_3 = \|\mathbf{X}^{(3)}\|$ are Rayleigh RVs corresponding to independent Gaussian vectors $\mathbf{X}^{(1)} \in N_n(\mathbf{0},\sigma_1^2)$ and $\mathbf{X}^{(3)} \in N_n(\mathbf{0},\sigma_1^2)$. Likewise consider the ratio RV $U_2 = R_4 / R_2$ where $R_2 = \|\mathbf{X}^{(2)}\|$ and $R_4 = \|\mathbf{X}^{(4)}\|$ are Rayleigh RVs corresponding to independent Gaussian vectors $\mathbf{X}^{(2)} \in N_n(\mathbf{0},\sigma_2^2)$ and $\mathbf{X}^{(4)} \in N_n(\mathbf{0},\sigma_2^2)$. The Gaussian vectors $\mathbf{X}^{(1)}$ and $\mathbf{X}^{(2)}$ are dependent and likewise the Gaussian vectors $\mathbf{X}^{(3)}$ and $\mathbf{X}^{(4)}$ are dependent with correlation matrix as in (1.10). Then U_1 and U_2 have the joint PDF.

$$
p_{U_1,U_2}(u_1,u_2) = \frac{4(1-\rho^2)^{2n}(u_1 u_2)^{n-1}}{\Gamma^2(n/2)(1+u_1^2)^n(1+u_2^2)^n}
$$

$$
\times \sum_{i=0}^{\infty}\sum_{l=0}^{\infty}\frac{[(n+i+l-1)!]^2}{i!\,l!\,\Gamma(i+n/2)\Gamma(l+n/2)}(u_1 u_2)^{2l}
\tag{7.92}
$$

$$
\times\left[\frac{\rho^2}{(1+u_1^2)(1+u_2^2)}\right]^{i+l},\ u_1,u_2\geq 0
$$

8

MAXIMUM AND MINIMUM OF PAIRS OF RANDOM VARIABLES

Finding the probability distributions of the maximum and minimum of groups of RVs involves the study of order statistics. Since a complete study of this subject on its own fills an entire textbook, examples of which are Refs. 12 and 13, we shall restrict our attention to the case of two RVs, in particular those who first and second-order probability distributions are discussed in Sections 2 and 3.

We denote the maximum and minimum of two RVs X_1 and X_2 by $X_{max} = \max(X_1, X_2)$ and $X_{min} = \min(X_1, X_2)$, respectively. Since $\Pr\{X_{max} \leq x\} = \Pr\{X_1 \leq x, X_2 \leq x\} = P_{X_1, X_2}(x, x)$, then the CDF of X_{max} is simply the joint CDF of X_1 and X_2 evaluated at the same argument, x, i.e.,

$$P_{X_{max}}(x) = P_{X_1, X_2}(x, x) \tag{8.1}$$

The PDF of X_{max} can then be obtained by differentiating the CDF in (8.1). Similarly, since $\Pr\{X_{min} > x\} = \Pr\{X_1 > x, X_2 > x\} = 1 - P_{X_1}(x) - P_{X_2}(x) + P_{X_1, X_2}(x, x)$, then the CDF of X_{min}, which is equal to $1 - \Pr\{X_{min} > x\}$, is given by

$$P_{X_{min}}(x) = P_{X_1}(x) + P_{X_2}(x) - P_{X_1, X_2}(x, x) \tag{8.2}$$

Again, the PDF of X_{min} can then be obtained by differentiating the CDF in (8.2).

For X_1 and X_2 independent, we have

$$P_{X_{max}}(x) = P_{X_1}(x) P_{X_2}(x) \tag{8.3}$$

$$p_{X_{max}}(x) = p_{X_1}(x) P_{X_2}(x) + p_{X_2}(x) P_{X_1}(x) \tag{8.4}$$

$$P_{X_{min}}(x) = P_{X_1}(x) + P_{X_2}(x) - P_{X_1}(x)P_{X_2}(x) \tag{8.5}$$

$$p_{X_{min}}(x) = p_{X_1}(x)\big(1 - P_{X_2}(x)\big) + p_{X_2}(x)\big(1 - P_{X_1}(x)\big) \tag{8.6}$$

A. Independent Gaussian

Let $X_1 \in N_1(\overline{X}_1, \sigma_1^2)$ and $X_2 \in N_1(\overline{X}_2, \sigma_2^2)$ be independent Gaussian RVs. Then,

$$p_{X_{max}}(x) = \frac{1}{\sqrt{2\pi\sigma_1^2}} \exp\left[-\frac{(x-\overline{X}_1)^2}{2\sigma_1^2}\right]\left[1 - Q\left(\frac{x-\overline{X}_2}{\sigma_2}\right)\right]$$
$$+ \frac{1}{\sqrt{2\pi\sigma_2^2}} \exp\left[-\frac{(x-\overline{X}_2)^2}{2\sigma_2^2}\right]\left[1 - Q\left(\frac{x-\overline{X}_1}{\sigma_1}\right)\right] \tag{8.7}$$

$$P_{X_{max}}(x) = \left[1 - Q\left(\frac{x-\overline{X}_1}{\sigma_1}\right)\right]\left[1 - Q\left(\frac{x-\overline{X}_2}{\sigma_2}\right)\right] \tag{8.8}$$

$$p_{X_{min}}(x) = \frac{1}{\sqrt{2\pi\sigma_1^2}} \exp\left[-\frac{(x-\overline{X}_1)^2}{2\sigma_1^2}\right]Q\left(\frac{x-\overline{X}_2}{\sigma_2}\right)$$
$$+ \frac{1}{\sqrt{2\pi\sigma_2^2}} \exp\left[-\frac{(x-\overline{X}_2)^2}{2\sigma_2^2}\right]Q\left(\frac{x-\overline{X}_1}{\sigma_1}\right) \tag{8.9}$$

$$P_{X_{min}}(x) = 1 - Q\left(\frac{x-\overline{X}_1}{\sigma_1}\right)Q\left(\frac{x-\overline{X}_2}{\sigma_2}\right) \tag{8.10}$$

B. Dependent Gaussian

$$p_{X_{max}}(x) = \frac{1}{\sqrt{2\pi\sigma_1^2}} \exp\left[-\frac{(x-\overline{X}_1)^2}{2\sigma_1^2}\right]\left\{1 - Q\left[\frac{\sigma_1(x-\overline{X}_2) - \rho\sigma_2(x-\overline{X}_1)}{\sigma_1\sigma_2\sqrt{1-\rho^2}}\right]\right\}$$
$$+ \frac{1}{\sqrt{2\pi\sigma_2^2}} \exp\left[-\frac{(x-\overline{X}_2)^2}{2\sigma_2^2}\right]\left\{1 - Q\left[\frac{\sigma_2(x-\overline{X}_1) - \rho\sigma_1(x-\overline{X}_2)}{\sigma_1\sigma_2\sqrt{1-\rho^2}}\right]\right\} \tag{8.11}$$

$$P_{X_{max}}(x) = 1 - Q\left(\frac{x-\bar{X}_1}{\sigma_1}\right) - Q\left(\frac{x-\bar{X}_2}{\sigma_2}\right) + Q_{X_1,X_2}\left(\frac{x-\bar{X}_1}{\sigma_1},\frac{x-\bar{X}_2}{\sigma_2};\rho\right) \quad (8.12)$$

(see Appendix A, Eq. (A.37) for definition of $Q_{X_1,X_2}(x,y;\rho)$).

$$p_{X_{min}}(x) = \frac{1}{\sqrt{2\pi\sigma_1^2}}\exp\left[-\frac{(x-\bar{X}_1)^2}{2\sigma_1^2}\right]Q\left[\frac{\sigma_1(x-\bar{X}_2)-\rho\sigma_2(x-\bar{X}_1)}{\sigma_1\sigma_2\sqrt{1-\rho^2}}\right]$$

$$+\frac{1}{\sqrt{2\pi\sigma_2^2}}\exp\left[-\frac{(x-\bar{X}_2)^2}{2\sigma_2^2}\right]Q\left[\frac{\sigma_2(x-\bar{X}_1)-\rho\sigma_1(x-\bar{X}_2)}{\sigma_1\sigma_2\sqrt{1-\rho^2}}\right] \quad (8.13)$$

$$P_{X_{min}}(x) = 1 - Q_{X_1,X_2}\left(\frac{x-\bar{X}_1}{\sigma_1},\frac{x-\bar{X}_2}{\sigma_2};\rho\right) \quad (8.14)$$

C. Independent Rayleigh

Let $R_1 = \|\mathbf{X}^{(1)}\|$ and $R_2 = \|\mathbf{X}^{(2)}\|$ be independent Rayleigh RVs corresponding to Gaussian vectors $\mathbf{X}^{(1)} \in N_n(0,\sigma_1^2)$ and $\mathbf{X}^{(2)} \in N_n(0,\sigma_2^2)$. Then, the maximum $R_{max} = \max(R_1, R_2)$ and minimum $R_{min} = \min(R_1, R_2)$ of the pair R_1, R_2 have the following statistical properties.

1. $n = 2$

$$p_{R_{max}}(r) = \frac{r}{\sigma_1^2}\exp\left(-\frac{r^2}{2\sigma_1^2}\right) + \frac{r}{\sigma_2^2}\exp\left(-\frac{r^2}{2\sigma_2^2}\right)$$

$$-\frac{r(\sigma_1^2+\sigma_2^2)}{\sigma_1^2\sigma_2^2}\exp\left[-\frac{r^2(\sigma_1^2+\sigma_2^2)}{2\sigma_1^2\sigma_2^2}\right], r \geq 0 \quad (8.15)$$

$$P_{R_{max}}(r) = \left[1-\exp\left(-\frac{r^2}{2\sigma_2^2}\right)\right]\left[1-\exp\left(-\frac{r^2}{2\sigma_1^2}\right)\right], r \geq 0 \quad (8.16)$$

$$p_{R_{min}}(r) = \frac{r(\sigma_1^2+\sigma_2^2)}{\sigma_1^2\sigma_2^2}\exp\left[-\frac{r^2(\sigma_1^2+\sigma_2^2)}{2\sigma_1^2\sigma_2^2}\right], r \geq 0 \quad (8.17)$$

$$P_{R_{min}}(r) = 1 - \exp\left[-\frac{r^2\left(\sigma_1^2 + \sigma_2^2\right)}{2\sigma_1^2\sigma_2^2}\right], \quad r \geq 0 \tag{8.18}$$

$$E\{R_{max}^k\} = \left[\left(2\sigma_1^2\right)^{k/2} + \left(2\sigma_1^2\right)^{k/2} - \left(\frac{2\sigma_1^2\sigma_2^2}{\sigma_1^2 + \sigma_2^2}\right)^{k/2}\right]\Gamma\left(1 + \frac{k}{2}\right), \quad k \text{ integer} \tag{8.19}$$

$$E\{R_{min}^k\} = \left(\frac{2\sigma_1^2\sigma_2^2}{\sigma_1^2 + \sigma_2^2}\right)^{k/2}\Gamma\left(1 + \frac{k}{2}\right), \quad k \text{ integer} \tag{8.20}$$

2. $n = 2m$

$$
\begin{aligned}
p_{R_{max}}(r) = {} & \frac{r}{\sigma_1^2(m-1)!}\left(\frac{r^2}{2\sigma_1^2}\right)^{m-1}\exp\left(-\frac{r^2}{2\sigma_1^2}\right)\left[1 - \exp\left(-\frac{r^2}{2\sigma_2^2}\right)\sum_{i=0}^{m-1}\frac{1}{i!}\left(\frac{r^2}{2\sigma_2^2}\right)^i\right] \\
& + \frac{r}{\sigma_2^2(m-1)!}\left(\frac{r^2}{2\sigma_2^2}\right)^{m-1}\exp\left(-\frac{r^2}{2\sigma_2^2}\right)\left[1 - \exp\left(-\frac{r^2}{2\sigma_1^2}\right)\sum_{i=0}^{m-1}\frac{1}{i!}\left(\frac{r^2}{2\sigma_1^2}\right)^i\right], \quad r \geq 0
\end{aligned}
\tag{8.21}
$$

$$
\begin{aligned}
P_{R_{max}}(r) = {} & 2 - \exp\left(-\frac{r^2}{2\sigma_1^2}\right)\sum_{i=0}^{m-1}\frac{1}{i!}\left(\frac{r^2}{2\sigma_1^2}\right)^i - \exp\left(-\frac{r^2}{2\sigma_2^2}\right)\sum_{i=0}^{m-1}\frac{1}{i!}\left(\frac{r^2}{2\sigma_2^2}\right)^i \\
& - \sum_{i=0}^{m-1}\binom{i+m-1}{i}\frac{\sigma_1^{2i}\sigma_2^{2m} + \sigma_1^{2m}\sigma_2^{2i}}{\left(\sigma_1^2 + \sigma_2^2\right)^{i+m}} \\
& \times \left\{1 - \exp\left[-\frac{r^2\left(\sigma_1^2 + \sigma_2^2\right)}{2\sigma_1^2\sigma_2^2}\right]\sum_{l=0}^{m+i-1}\frac{1}{l!}\left[\frac{r^2\left(\sigma_1^2 + \sigma_2^2\right)}{2\sigma_1^2\sigma_2^2}\right]^l\right\}, \quad r \geq 0
\end{aligned}
\tag{8.22}
$$

$$
\begin{aligned}
p_{R_{min}}(r) = {} & \frac{r}{\sigma_1^2(m-1)!}\left(\frac{r^2}{2\sigma_1^2}\right)^{m-1}\exp\left[-\frac{r^2\left(\sigma_1^2 + \sigma_2^2\right)}{2\sigma_1^2\sigma_2^2}\right]\sum_{i=0}^{m-1}\frac{1}{i!}\left(\frac{r^2}{2\sigma_2^2}\right)^i \\
& + \frac{r}{\sigma_2^2(m-1)!}\left(\frac{r^2}{2\sigma_2^2}\right)^{m-1}\exp\left[-\frac{r^2\left(\sigma_1^2 + \sigma_2^2\right)}{2\sigma_1^2\sigma_2^2}\right]\sum_{i=0}^{m-1}\frac{1}{i!}\left(\frac{r^2}{2\sigma_1^2}\right)^i, \quad r \geq 0
\end{aligned}
\tag{8.23}
$$

$$P_{R_{min}}(r) = \sum_{i=0}^{m-1} \binom{i+m-1}{i} \frac{\sigma_1^{2i}\sigma_2^{2m} + \sigma_1^{2m}\sigma_2^{2i}}{(\sigma_1^2 + \sigma_2^2)^{i+m}}$$

$$\times \left\{ 1 - \exp\left[-\frac{r^2(\sigma_1^2 + \sigma_2^2)}{2\sigma_1^2\sigma_2^2} \right]^{m+i-1} \sum_{l=0}^{1} \frac{1}{l!} \left[\frac{r^2(\sigma_1^2 + \sigma_2^2)}{2\sigma_1^2\sigma_2^2} \right]^l \right\}, r \geq 0 \tag{8.24}$$

D. Dependent Rayleigh

1. $n = 2$

$$P_{R_{max}}(r) = \frac{r}{\sigma_1^2} \exp\left(-\frac{r^2}{2\sigma_1^2} \right) \left[1 - Q_1\left(\sqrt{\frac{\rho^2 r^2}{(1-\rho^2)\sigma_1^2}}, \sqrt{\frac{r^2}{(1-\rho^2)\sigma_2^2}} \right) \right]$$

$$+ \frac{r}{\sigma_2^2} \exp\left(-\frac{r^2}{2\sigma_2^2} \right) \left[1 - Q_1\left(\sqrt{\frac{\rho^2 r^2}{(1-\rho^2)\sigma_2^2}}, \sqrt{\frac{r^2}{(1-\rho^2)\sigma_1^2}} \right) \right], r \geq 0 \tag{8.25}$$

$$P_{R_{max}}(r) = G\big(f_1(r,\sigma_1^2), f_1(r,\sigma_2^2)\big) - \frac{1}{2\pi} \int_{-\pi}^{\pi} h_1^{-1}(\theta) f_1(r, h_1^{-1}(\theta)) h(\theta) d\theta, r \geq 0 \tag{8.26}$$

$$P_{R_{min}}(r) = \frac{r}{\sigma_1^2} \exp\left(-\frac{r^2}{2\sigma_1^2} \right) Q_1\left(\sqrt{\frac{\rho^2 r^2}{(1-\rho^2)\sigma_1^2}}, \sqrt{\frac{r^2}{(1-\rho^2)\sigma_2^2}} \right)$$

$$+ \frac{r}{\sigma_2^2} \exp\left(-\frac{r^2}{2\sigma_2^2} \right) Q_1\left(\sqrt{\frac{\rho^2 r^2}{(1-\rho^2)\sigma_2^2}}, \sqrt{\frac{r^2}{(1-\rho^2)\sigma_1^2}} \right), r \geq 0 \tag{8.27}$$

$$P_{R_{min}}(r) = f_1(r,\sigma_1^2) + f_1(r,\sigma_2^2) - G\big(f_1(r,\sigma_1^2), f_1(r,\sigma_2^2)\big)$$

$$+ \frac{1}{2\pi} \int_{-\pi}^{\pi} h_1^{-1}(\theta) f_1(r, h_1^{-1}(\theta)) h(\theta) d\theta, r \geq 0 \tag{8.28}$$

where

$$f_m(r,\eta) \triangleq 1 - \exp\left(-\frac{r^2}{2\eta} \right) \sum_{i=0}^{m-1} \frac{1}{i!} \left(\frac{r^2}{2\eta} \right)^i \tag{8.29}$$

$$h(\theta) \triangleq \frac{1}{\left(2\sigma_1^2\right)^m}\left(\frac{\sigma_1^2}{\rho^2\sigma_2^2}\right)^{(m-1)/2}\left[\frac{\begin{aligned}-\sigma_1^2\cos\left[(m-1)(\theta+\pi/2)\right] \\ +|\rho|\sigma_1\sigma_2\cos\left[m(\theta+\pi/2)\right]\end{aligned}}{\rho^2\sigma_2^2+2|\rho|\sigma_1\sigma_2\sin\theta+\sigma_1^2}\right]$$

$$+\frac{1}{\left(2\sigma_1^2\right)^m}\left(\frac{\rho^2\sigma_1^2}{\sigma_2^2}\right)^{-(m-1)/2}\left[\frac{\begin{aligned}\sigma_2^2\cos\left[(m-1)(\theta+\pi/2)\right] \\ -|\rho|\sigma_1\sigma_2\cos\left[m(\theta+\pi/2)\right]\end{aligned}}{\sigma_2^2+2|\rho|\sigma_1\sigma_2\sin\theta+\rho^2\sigma_1^2}\right]$$

$$(8.30)$$

$$h_1(\theta) \triangleq \frac{\sigma_1^2+\sigma_2^2+2|\rho|\sigma_1\sigma_2\sin\theta}{2\sigma_1^2\sigma_2^2\left(1-\rho^2\right)} \tag{8.31}$$

and

$$G\left(f_m\left(r,\sigma_1^2\right),f_m\left(r,\sigma_2^2\right)\right)\triangleq\begin{cases} f_m\left(r,\sigma_2^2\right), & \sigma_1^2<\rho^2\sigma_2^2 \\[2mm] \dfrac{1}{2}f_m\left(r,\sigma_1^2\right)+f_m\left(r,\sigma_2^2\right), & \sigma_1^2=\rho^2\sigma_2^2 \\[2mm] f_m\left(r,\sigma_1^2\right)+f_m\left(r,\sigma_2^2\right), & \rho^2\sigma_2^2<\sigma_1^2<\dfrac{1}{\rho^2}\sigma_2^2 \\[2mm] f_m\left(r,\sigma_1^2\right)+\dfrac{1}{2}f_m\left(r,\sigma_2^2\right), & \sigma_1^2=\dfrac{1}{\rho^2}\sigma_2^2 \\[2mm] f_m\left(r,\sigma_1^2\right), & \dfrac{1}{\rho^2}\sigma_2^2<\sigma_1^2 \end{cases} \tag{8.32}$$

2. $n=2m$

$$p_{R_{\max}}(r)=\frac{r}{\sigma_1^2(m-1)!}\left(\frac{r^2}{2\sigma_1^2}\right)^{m-1}\exp\left(-\frac{r^2}{2\sigma_1^2}\right)\left[1-Q_m\left(\sqrt{\frac{\rho^2 r^2}{\left(1-\rho^2\right)\sigma_1^2}},\sqrt{\frac{r^2}{\left(1-\rho^2\right)\sigma_2^2}}\right)\right]$$

$$+\frac{r}{\sigma_2^2(m-1)!}\left(\frac{r^2}{2\sigma_2^2}\right)^{m-1}\exp\left(-\frac{r^2}{2\sigma_2^2}\right)\left[1-Q_m\left(\sqrt{\frac{\rho^2 r^2}{\left(1-\rho^2\right)\sigma_2^2}},\sqrt{\frac{r^2}{\left(1-\rho^2\right)\sigma_1^2}}\right)\right],$$

$$r\geq 0$$

$$P_{R_{max}}(r) = G\big(f_m(r,\sigma_1^2), f_m(r,\sigma_2^2)\big) \tag{8.33}$$

$$-\frac{m^m}{(m-1)!}\frac{1}{2\pi}\int_{-\pi}^{\pi}\big(h_1(\theta)\big)^{-m}f_m(r,h_1^{-1}(\theta))h(\theta)d\theta, \; r\geq 0 \tag{8.34}$$

$$P_{R_{min}}(r) = \frac{r}{\sigma_1^2(m-1)!}\left(\frac{r^2}{2\sigma_1^2}\right)^{m-1}\exp\left(-\frac{r^2}{2\sigma_1^2}\right)$$

$$\times Q_m\left(\sqrt{\frac{\rho^2 r^2}{(1-\rho^2)\sigma_1^2}}, \sqrt{\frac{r^2}{(1-\rho^2)\sigma_2^2}}\right)$$

$$+\frac{r}{\sigma_2^2(m-1)!}\left(\frac{r^2}{2\sigma_2^2}\right)^{m-1}\exp\left(-\frac{r^2}{2\sigma_2^2}\right) \tag{8.35}$$

$$\times Q_m\left(\sqrt{\frac{\rho^2 r^2}{(1-\rho^2)\sigma_2^2}}, \sqrt{\frac{r^2}{(1-\rho^2)\sigma_1^2}}\right), \; r\geq 0$$

$$P_{R_{min}}(r) = f_m(r,\sigma_1^2) + f_m(r,\sigma_2^2) - G\big(f_m(r,\sigma_1^2), f_m(r,\sigma_2^2)\big)$$

$$+\frac{m^m}{(m-1)!}\frac{1}{2\pi}\int_{-\pi}^{\pi}\big(h_1(\theta)\big)^{-m}f_m(r,h_1^{-1}(\theta))h(\theta)d\theta, \; r\geq 0 \tag{8.36}$$

E. Independent Log-Normal

Let $\gamma_1 = 10^{X_1/10}$ and $\gamma_2 = 10^{X_2/10}$ be independent log-normal RVs corresponding to Gaussian RVs $X_1 \in N_1(\overline{X}_1,\sigma^2)$ and $X_2 \in N_1(\overline{X}_2,\sigma^2)$. Then, the maximum $\gamma_{max} = \max(\gamma_1,\gamma_2)$ and minimum $\gamma_{min} = \min(\gamma_1,\gamma_2)$ of the pair γ_1,γ_2 have the following statistical properties.

$$P_{\gamma_{max}}(\gamma) = \frac{\xi}{\sqrt{2\pi}\sigma_1\gamma}\exp\left[-\frac{(10\log_{10}\gamma - \overline{X}_1)^2}{2\sigma_1^2}\right]\left[1 - Q\left(\frac{10\log_{10}\gamma - \overline{X}_2}{\sigma_2}\right)\right]$$

$$+\frac{\xi}{\sqrt{2\pi}\sigma_2\gamma}\exp\left[-\frac{(10\log_{10}\gamma - \overline{X}_2)^2}{2\sigma_2^2}\right]\left[1 - Q\left(\frac{10\log_{10}\gamma - \overline{X}_1}{\sigma_1}\right)\right], \gamma\geq 0$$

$$\tag{8.37}$$

$$P_{\gamma_{max}}(\gamma) = \left[1 - Q\left(\frac{10\log_{10}\gamma - \overline{X}_1}{\sigma_1}\right)\right]\left[1 - Q\left(\frac{10\log_{10}\gamma - \overline{X}_2}{\sigma_2}\right)\right], \gamma \geq 0 \quad (8.38)$$

$$p_{\gamma_{min}}(\gamma) = \frac{\xi}{\sqrt{2\pi}\sigma_1\gamma}\exp\left[-\frac{(10\log_{10}\gamma - \overline{X}_1)^2}{2\sigma_1^2}\right]Q\left(\frac{10\log_{10}\gamma - \overline{X}_2}{\sigma_2}\right)$$

$$+\frac{\xi}{\sqrt{2\pi}\sigma_2\gamma}\exp\left[-\frac{(10\log_{10}\gamma - \overline{X}_2)^2}{2\sigma_2^2}\right]Q\left(\frac{10\log_{10}\gamma - \overline{X}_1}{\sigma_1}\right), \gamma \geq 0 \quad (8.39)$$

$$P_{\gamma_{min}}(\gamma) = 1 - Q\left(\frac{10\log_{10}\gamma - \overline{X}_1}{\sigma_1}\right)Q\left(\frac{10\log_{10}\gamma - \overline{X}_2}{\sigma_2}\right), \gamma \geq 0 \quad (8.40)$$

$$E\{\gamma_{max}^k\} = \exp\left\{\frac{k\overline{X}_1}{\xi} + \frac{k^2\sigma_1^2}{2\xi^2}\right\}Q\left(\frac{\overline{X}_2 - \overline{X}_1 - \dfrac{k\sigma_1^2}{\xi}}{\sqrt{\sigma_1^2 + \sigma_2^2}}\right)$$

$$+\exp\left\{\frac{k\overline{X}_2}{\xi} + \frac{k^2\sigma_2^2}{2\xi^2}\right\}Q\left(\frac{\overline{X}_1 - \overline{X}_2 - \dfrac{k\sigma_2^2}{\xi}}{\sqrt{\sigma_1^2 + \sigma_2^2}}\right) \quad (8.41)$$

F. Dependent Log-Normal

$$p_{\gamma_{max}}(\gamma) = \frac{\xi}{\sqrt{2\pi}\sigma_1\gamma}\exp\left[-\frac{(10\log_{10}\gamma - \overline{X}_1)^2}{2\sigma_1^2}\right]$$

$$\times\left\{1 - Q\left[\frac{\sigma_1(10\log_{10}\gamma - \overline{X}_2) - \rho\sigma_2(10\log_{10}\gamma - \overline{X}_1)}{\sigma_1\sigma_2\sqrt{1-\rho^2}}\right]\right\}$$

$$+\frac{\xi}{\sqrt{2\pi}\sigma_2\gamma}\exp\left[-\frac{(10\log_{10}\gamma - \overline{X}_2)^2}{2\sigma_2^2}\right]$$

$$\times\left\{1 - Q\left[\frac{\sigma_2(10\log_{10}\gamma - \overline{X}_1) - \rho\sigma_1(10\log_{10}\gamma - \overline{X}_2)}{\sigma_1\sigma_2\sqrt{1-\rho^2}}\right]\right\}, \gamma \geq 0 \quad (8.42)$$

$$P_{\gamma_{max}}(\gamma) = 1 - Q\left(\frac{10\log_{10}\gamma - \overline{X}_1}{\sigma_1}\right) - Q\left(\frac{10\log_{10}\gamma - \overline{X}_2}{\sigma_2}\right)$$

$$+ Q\left(\frac{10\log_{10}\gamma - \overline{X}_1}{\sigma_1}, \frac{10\log_{10}\gamma - \overline{X}_2}{\sigma_2}; \rho\right), \gamma \geq 0 \qquad (8.43)$$

$$p_{\gamma_{min}}(\gamma) = \frac{\xi}{\sqrt{2\pi}\sigma_1\gamma}\exp\left[-\frac{\left(10\log_{10}\gamma - \overline{X}_1\right)^2}{2\sigma_1^2}\right]$$

$$\times Q\left[\frac{\sigma_1\left(10\log_{10}\gamma - \overline{X}_2\right) - \rho\sigma_2\left(10\log_{10}\gamma - \overline{X}_1\right)}{\sigma_1\sigma_2\sqrt{1-\rho^2}}\right]$$

$$+ \frac{\xi}{\sqrt{2\pi}\sigma_2\gamma}\exp\left[-\frac{\left(10\log_{10}\gamma - \overline{X}_2\right)^2}{2\sigma_2^2}\right] \qquad (8.44)$$

$$\times Q\left[\frac{\sigma_2\left(10\log_{10}\gamma - \overline{X}_1\right) - \rho\sigma_1\left(10\log_{10}\gamma - \overline{X}_2\right)}{\sigma_1\sigma_2\sqrt{1-\rho^2}}\right], \gamma \geq 0$$

$$P_{\gamma_{min}}(\gamma) = 1 - Q\left(\frac{10\log_{10}\gamma - \overline{X}_1}{\sigma_1}, \frac{10\log_{10}\gamma - \overline{X}_2}{\sigma_2}; \rho\right), \gamma \geq 0 \qquad (8.45)$$

$$E\{\gamma_{max}^k\} = \exp\left\{\frac{k\overline{X}_1}{\xi} + \frac{k^2\sigma_1^2}{2\xi^2}\right\}Q\left(\frac{\overline{X}_2 - \overline{X}_1 - \dfrac{k\sigma_1^2}{\xi}\left(1 - \dfrac{\rho\sigma_2}{\sigma_1}\right)}{\sqrt{\sigma_1^2 + \sigma_2^2 - 2\rho\sigma_1\sigma_2}}\right)$$

$$+ \exp\left\{\frac{k\overline{X}_2}{\xi} + \frac{k^2\sigma_2^2}{2\xi^2}\right\}Q\left(\frac{\overline{X}_1 - \overline{X}_2 - \dfrac{k\sigma_2^2}{\xi}\left(1 - \dfrac{\rho\sigma_1}{\sigma_2}\right)}{\sqrt{\sigma_1^2 + \sigma_2^2}}\right) \qquad (8.46)$$

9

QUADRATIC FORMS

Let $\mathbf{X}$ and $\mathbf{Y}$ be complex Gaussian vectors each of dimension n. As usual, the components of each vector are independent and have identical variances, which for complex components are defined as

$$\mu_{xx} = \frac{1}{2}E\left\{\left|X_k - \bar{X}_k\right|^2\right\}, \quad \mu_{yy} = \frac{1}{2}E\left\{\left|Y_k - \bar{Y}_k\right|^2\right\} \tag{9.1}$$

Furthermore, as before only components of $\mathbf{X}$ and $\mathbf{Y}$ with identical subscripts can be correlated with complex cross-correlation defined by

$$\mu_{xy} = \frac{1}{2}E\left\{\left(X_k - \bar{X}_k\right)\left(Y_k - \bar{Y}_k\right)^*\right\} \tag{9.2}$$

For convenience, we define pairs of complex components with the same subscript by the vectors

$$\mathbf{V}_k = \begin{bmatrix} X_k \\ Y_k \end{bmatrix}, \quad k = 1, 2, \ldots, n \tag{9.3}$$

and analogous to (1.10) we define the covariance matrix

$$\mathbf{M} = \begin{bmatrix} \mu_{xx} & \mu_{xy}^* \\ \mu_{xy} & \mu_{yy} \end{bmatrix} \tag{9.4}$$

We are interested in the statistics of the RV having the quadratic form

$$D = \sum_{k=1}^{n} d_k \tag{9.5}$$

where

$$d_k = A|X_k|^2 + B|Y_k|^2 + CX_kY_k^* + C^*X_k^*Y_k \tag{9.6}$$

which can also be written in the vector form

$$d_k = V_k^{*T}HV_k, \quad H = \begin{bmatrix} A & C^* \\ C & B \end{bmatrix} \tag{9.7}$$

Finally, we define the matrix **P** by

$$P = MH = \begin{bmatrix} p_{xx} & p_{xy} \\ p_{yx} & p_{yy} \end{bmatrix} \tag{9.8}$$

where from (9.4) and (9.7) we obtain

$$p_{xx} = A\mu_{xx} + C\mu_{xy}^*, \quad p_{yy} = B\mu_{yy} + C^*\mu_{xy}$$
$$p_{xy} = C^*\mu_{xx} + B\mu_{xy}^*, \quad p_{yx} = C\mu_{yy} + A\mu_{xy} \tag{9.9}$$

Also, for what follows it is convenient to define the parameters

$$v_1 = \left\{ \left[\frac{p_{xx} + p_{yy}}{4|P|} \right]^2 - \frac{1}{4|P|} \right\}^{1/2} + \frac{p_{xx} + p_{yy}}{4|P|}$$

$$v_2 = \left\{ \left[\frac{p_{xx} + p_{yy}}{4|P|} \right]^2 - \frac{1}{4|P|} \right\}^{1/2} - \frac{p_{xx} + p_{yy}}{4|P|} \tag{9.10}$$

which when using (9.9) evaluate to

$$v_1 = \sqrt{w^2 + \frac{1}{4\left(\mu_{xx}\mu_{yy} - |\mu_{xy}|^2\right)\left(|C|^2 - AB\right)}} - w$$

$$v_2 = \sqrt{w^2 + \frac{1}{4\left(\mu_{xx}\mu_{yy} - |\mu_{xy}|^2\right)\left(|C|^2 - AB\right)}} + w \tag{9.11}$$

with

$$w = \frac{A\mu_{xx} + B\mu_{yy} + C\mu_{xy}^{*} + C^{*}\mu_{xy}}{4\left(\mu_{xx}\mu_{yy} - \left|\mu_{xy}\right|^{2}\right)\left(\left|C\right|^{2} - AB\right)} \tag{9.12}$$

A. Both Vectors Have Zero Mean

1. $n = 1$

$$p_D(d) = \begin{cases} \dfrac{v_1 v_2}{v_1 + v_2} \exp(v_2 d), & d < 0 \\[3mm] \dfrac{v_1 v_2}{v_1 + v_2} \exp(-v_1 d), & d \ge 0 \end{cases} \tag{9.13}$$

$$P_D(d) = \begin{cases} \dfrac{v_1}{v_1 + v_2} \exp(v_2 d), & d < 0 \\[3mm] 1 - \dfrac{v_2}{v_1 + v_2} \exp(-v_1 d), & d \ge 0 \end{cases} \tag{9.14}$$

2. n

$$p_D(d) = \begin{cases} \exp(v_2 d) \displaystyle\sum_{k=0}^{n-1} \frac{(v_1 v_2)^n}{(v_1 + v_2)^{n+k}} \binom{n+k-1}{k} \frac{(-d)^{n-k-1}}{(n-k-1)!}, & d < 0 \\[5mm] \exp(-v_1 d) \displaystyle\sum_{k=0}^{n-1} \frac{(v_1 v_2)^n}{(v_1 + v_2)^{n+k}} \binom{n+k-1}{k} \frac{(-d)^{n-k-1}}{(n-k-1)!}, & d \ge 0 \end{cases} \tag{9.15}$$

$$P_D(d) = \begin{cases} \exp(v_2 d) \displaystyle\sum_{k=0}^{n-1}\sum_{i=0}^{n-k-1} \frac{(v_1 v_2)^n}{(v_1 + v_2)^{n+k}} \binom{n+k-1}{k} \frac{(-d)^{n-k-i-1}}{(n-k-i-1)!\, v_2^{i+1}}, & d < 0 \\[5mm] 1 - \exp(-v_1 d) \displaystyle\sum_{k=0}^{n-1} \frac{(v_1 v_2)^n}{(v_1 + v_2)^{n+k}} \binom{n+k-1}{k} \frac{(-1)^{n-k-1}(-d)^{n-k-i-1}}{(n-k-i-1)!\, v_1^{i+1}}, & d \ge 0 \end{cases} \tag{9.16}$$

A special case of (9.16) that is of practical interest (see [14], Appendix B and [15], Appendix 9A, for example as applied to the study of the performance of digital communication systems over fading channels) corresponds to $P_D(0) = \Pr\{D \le 0\}$. For this case, (9.16) reduces to

$$P_D(0) = \sum_{k=0}^{n-1} \frac{v_1^n v_2^k}{\left(v_1 + v_2\right)^{n+k}} \binom{n+k-1}{k} \tag{9.17}$$

An alternative form for $P_D(0)$ which will appear as a special case of the results to be presented in the next section for nonzero mean vectors is given by

$$P_D(0) = \frac{\sum_{k=0}^{n-1} \binom{2n-1}{k}\left(\dfrac{v_2}{v_1}\right)^k}{\left(1+\dfrac{v_2}{v_1}\right)^{2n-1}} = \frac{\sum_{k=0}^{n-1} \binom{2n-1}{k} v_1^{2n-1-k} v_2^k}{\left(v_1 + v_2\right)^{2n-1}} \tag{9.18}$$

B. One or Both Vectors Have Nonzero Mean

When one or both of the two vectors have nonzero mean, finding closed-form expressions for the PDF and CDF of the quadratic form in (9.5) is difficult. However, for the particular case corresponding to $P_D(0)$, it is possible to obtain such expressions [14], [15]. Two different closed-forms versions will be provided, the first involving the first-order Marcum Q-function and modified Bessel functions of order $0,1,...,n-1$(as well as other elementary functions) and the second involving only the zero-order modified Bessel function but instead Marcum Q-functions of order $1,2,...,n$. Also provided will be a finite-limit integral form for $P_D(0)$ that is derived based on the alternative representation of the Marcum Q-function discussed in Appendix A and is convenient for numerical evaluation purposes. The results are as follows:

$$P_D(0) = Q_1(a,b) - I_0(ab)\exp\left[-\tfrac{1}{2}\left(a^2 + b^2\right)\right]$$

$$+ \frac{I_0(ab)\exp\left[-\tfrac{1}{2}\left(a^2 + b^2\right)\right]}{\left(1+v_2/v_1\right)^{2n-1}}\sum_{k=0}^{n-1}\binom{2n-1}{k}\left(\frac{v_2}{v_1}\right)^k + \frac{\exp\left[-\tfrac{1}{2}\left(a^2 + b^2\right)\right]}{\left(1+v_2/v_1\right)^{2n-1}} \tag{9.19}$$

$$\times \sum_{i=1}^{n-1} I_i(ab)\sum_{k=0}^{n-1-i}\binom{2n-1}{k}\left[\left(\frac{b}{a}\right)^i\left(\frac{v_2}{v_1}\right)^k - \left(\frac{a}{b}\right)^i\left(\frac{v_2}{v_1}\right)^{2n-1-k}\right]$$

where

$$a = \left[\frac{2v_1^2 v_2 (\xi_1 v_2 - \xi_2)}{(v_1 + v_2)^2} \right]^{1/2}, \quad b = \left[\frac{2v_1^2 v_2 (\xi_1 v_1 + \xi_2)}{(v_1 + v_2)^2} \right]^{1/2} \tag{9.20}$$

with

$$\xi_1 = \sum_{k=1}^{n} \xi_{1k}, \quad \xi_2 = \sum_{k=1}^{n} \xi_{2k}$$

$$\xi_{1k} = 2 \left(|C|^2 - AB \right) \left(|\bar{X}_k|^2 \mu_{yy} + |\bar{Y}_k|^2 \mu_{xx} - \bar{X}_k^* \bar{Y}_k \mu_{xy} - \bar{X}_k \bar{Y}_k^* \mu_{xy}^* \right), \tag{9.21}$$

$$\xi_{2k} = \left(A |\bar{X}_k|^2 + B |\bar{Y}_k|^2 + C \bar{X}_k^* \bar{Y}_k + C^* \bar{X}_k \bar{Y}_k^* \right)$$

$$P_D(0) = Q_1(a,b) - \left[1 - \frac{\sum_{k=0}^{n-1} \binom{2n-1}{k} \left(\frac{v_2}{v_1} \right)^k}{\left(1 + \frac{v_2}{v_1} \right)^{2n-1}} \right] \exp\left(-\frac{a^2 + b^2}{2} \right) I_0(ab)$$

$$+ \frac{1}{\left(1 + \frac{v_2}{v_1} \right)^{2n-1}} \left[\sum_{l=2}^{n} \binom{2n-1}{n-l} \left(\frac{v_2}{v_1} \right)^{n-l} \left[Q_l(a,b) - Q_1(a,b) \right] \right. \tag{9.22}$$

$$\left. - \sum_{l=2}^{n} \binom{2n-1}{n-l} \left(\frac{v_2}{v_1} \right)^{n-1+l} \left[Q_l(b,a) - Q_1(b,a) \right] \right]$$

$$P_D(0) = \frac{\left(\frac{v_2}{v_1} \right)^n}{2\pi \left(1 + \frac{v_2}{v_1} \right)^{2n-1}} \int_{-\pi}^{\pi} \left[\frac{b^2 f \left(n; \frac{a}{b}, \frac{v_2}{v_1}; \theta \right)}{a^2 + 2ab\sin\theta + b^2} \right]$$

$$\times \exp\left[-\frac{1}{2} \left(a^2 + 2ab\sin\theta + b^2 \right) \right] d\theta, \quad a < b \tag{9.23}$$

where

$$f(n; \zeta, \eta; \theta) = \sum_{l=1}^{n} \binom{2n-1}{n-l} \left[\left(\eta^{-l} \zeta^{-l+1} - \eta^{l-1} \zeta^{l+1} \right) \cos\left((l-1)(\theta + \tfrac{\pi}{2}) \right) \right.$$

$$\left. - \left(\eta^{-l} \zeta^{-l+2} - \eta^{l-1} \zeta^l \right) \cos\left(l(\theta + \tfrac{\pi}{2}) \right) \right] \tag{9.24}$$

Note that the form of $P_D(0)$ in (9.22) does not depend on the ratio a/b and thus for the case where both vectors have zero mean, whereupon $a = b = 0$, using the fact that $Q_i(0,0) = 1$, we immediately obtain the result in (9.18).

C. A Reduced Quadratic Form Where the Vectors Have Different Numbers of Dimensions

We consider now a reduced version of the quadratic form in (9.5) combined with (9.6), namely,

$$D = \sum_{k=1}^{n_1} A|X_k|^2 + \sum_{k=1}^{n_2} B|Y_k|^2 \tag{9.25}$$

where the complex Gaussian vectors $\mathbf{X}$ and $\mathbf{Y}$ now have dimensions n_1 and n_2, respectively. Once again for the most general case where one or both of the two vectors have nonzero mean, finding closed-form expressions for the PDF and CDF of the quadratic form in (9.25) is difficult. As in the previous section, however, it is possible to arrive at closed-form expressions for $P_D(0)$ [16]. Analogous to (9.22) and (9.23), the results are as follows:

$$
\begin{aligned}
P_D(0) = {} & Q_1(a,b) - \exp\left[-\tfrac{1}{2}\left(a^2 + b^2\right)\right] I_0(ab) \\
& + \frac{I_0(ab)\exp\left[-\tfrac{1}{2}\left(a^2 + b^2\right)\right]}{\left[1 + v_2 / v_1\right]^{L_1 + L_2 - 1}} \sum_{k=0}^{n_1 - 1} \binom{n_1 + n_2 - 1}{k} \left(\frac{v_2}{v_1}\right)^k \\
& + \frac{\exp\left[-\tfrac{1}{2}\left(a^2 + b^2\right)\right]}{\left[1 + v_2 / v_1\right]^{n_1 + n_2 - 1}} \left[\sum_{i=1}^{n_1 - 1} I_i(ab) \sum_{k=0}^{n_1 - 1 - i} \binom{n_1 + n_2 - 1}{k} \left(\frac{b}{a}\right)^i \left(\frac{v_2}{v_1}\right)^k \right. \\
& \left. - \sum_{i=1}^{n_2 - 1} I_n(ab) \sum_{k=0}^{n_2 - 1 - i} \binom{n_1 + n_2 - 1}{k} \left(\frac{a}{b}\right)^i \left(\frac{v_2}{v_1}\right)^{n_1 + n_2 - 1 - k} \right], \quad n_1, n_2 > 1
\end{aligned} \tag{9.26}
$$

$$P_D(0) = Q_1(a,b) - \left[1 - \frac{\displaystyle\sum_{k=0}^{n_1-1} \binom{n_1+n_2-1}{k}\left(\frac{v_2}{v_1}\right)^k}{\left(1+\dfrac{v_2}{v_1}\right)^{n_1+n_2-1}} \right] \exp\left(-\frac{a^2+b^2}{2}\right) I_0(ab)$$

$$+ \frac{1}{\left(1+\dfrac{v_2}{v_1}\right)^{n_1+n_2-1}} \left[\sum_{l=2}^{n_1} \binom{n_1+n_2-1}{n_1-l}\left(\frac{v_2}{v_1}\right)^{n_1-l} [Q_l(a,b) - Q_1(a,b)] \right.$$

$$\left. - \sum_{l=2}^{n_2} \binom{n_1+n_2-1}{n_2-l}\left(\frac{v_2}{v_1}\right)^{n_1-1+l} [Q_l(b,a) - Q_1(b,a)] \right] \qquad (9.27)$$

$$P_D(0) = \frac{\left(\dfrac{v_2}{v_1}\right)^{n_1}}{2\pi\left(1+\dfrac{v_2}{v_1}\right)^{n_1+n_2-1}} \int_{-\pi}^{\pi} \left[\frac{b^2 f\left(n_1,n_2;\dfrac{a}{b},\dfrac{v_2}{v_1};\theta\right)}{a^2 + 2ab\sin\theta + b^2} \right]$$

$$\times \exp\left[-\frac{1}{2}(a^2 + 2ab\sin\theta + b^2)\right] d\theta, \quad a < b \qquad (9.28)$$

where

$$f(n_1,n_2;\zeta,\eta;\theta) = \sum_{l=1}^{n_1}\binom{n_1+n_2-1}{n_1-l}\eta^{-l}\zeta^{-l+1}\left[\cos((l-1)(\theta+\tfrac{\pi}{2})) - \zeta\cos(l(\theta+\tfrac{\pi}{2}))\right]$$

$$+ \sum_{l=1}^{n_2}\binom{n_1+n_2-1}{n_2-l}\eta^{l-1}\zeta^l\left[\cos(l(\theta+\tfrac{\pi}{2})) - \zeta\cos((l-1)(\theta+\tfrac{\pi}{2}))\right]$$

$$(9.29)$$

and because $C = 0$, a and b of (9.20) simplify to

$$a = \sqrt{\frac{\displaystyle\sum_{k=1}^{n_1} A|\overline{X}_k|^2}{A\mu_{xx} - B\mu_{yy}}}, \quad b = \sqrt{\frac{-\displaystyle\sum_{k=1}^{n_2} B|\overline{Y}_k|^2}{A\mu_{xx} - B\mu_{yy}}} \qquad (9.30)$$

Note that since $\mu_{xx}, \mu_{yy} > 0$, then for either $A < 0$ or $B < 0$, the arguments of the square roots in (9.30) are always positive.

D. General Hermetian Quadratic Forms

As an alternative to describing each component of the quadratic form of (9.5) in a matrix form as in (9.7), the quadratic form itself can be expressed in a matrix form as follows. Let

$$
V = \begin{bmatrix} V_1 \\ V_2 \\ . \\ . \\ V_n \end{bmatrix} = \begin{bmatrix} X_1 \\ Y_1 \\ X_2 \\ Y_2 \\ . \\ . \\ X_n \\ Y_n \end{bmatrix}, \quad Q = \begin{bmatrix} H & 0 & . & . & 0 \\ 0 & H & 0 & . & 0 \\ . & 0 & . & . & . \\ . & . & . & . & 0 \\ 0 & 0 & . & 0 & H \end{bmatrix} \tag{9.31}
$$

Then, analogous to (9.7),

$$
D = V^{*T} Q V \tag{9.32}
$$

Note from (9.31) together with (9.7) that the matrix Q is Hermetian, i.e., $Q = Q^{*T}$.

Consider now a general quadratic form of the type in (9.32) where Q is again Hermetian but not restricted to the diagonal form in (9.31). Furthermore, denoting the complex elements of V by $X_n = X_{nR} + jX_{nI}$ and $Y_n = Y_{nR} + jY_{nI}$, assume that their real and imaginary parts satisfy the relations

$$
\begin{aligned}
E\left\{\left(X_{mR} - \bar{X}_{mR}\right)\left(X_{nR} - \bar{X}_{nR}\right)\right\} &= E\left\{\left(X_{mI} - \bar{X}_{mI}\right)\left(X_{nI} - \bar{X}_{nI}\right)\right\} \\
E\left\{\left(X_{mR} - \bar{X}_{mR}\right)\left(X_{nI} - \bar{X}_{nI}\right)\right\} &= -E\left\{\left(X_{nI} - \bar{X}_{nI}\right)\left(X_{mI} - \bar{X}_{mI}\right)\right\}
\end{aligned}
\tag{9.33}
$$

and

$$
\begin{aligned}
E\left\{\left(Y_{mR} - \bar{Y}_{mR}\right)\left(Y_{nR} - \bar{Y}_{nR}\right)\right\} &= E\left\{\left(Y_{mI} - \bar{Y}_{mI}\right)\left(Y_{nI} - \bar{Y}_{nI}\right)\right\} \\
E\left\{\left(Y_{mR} - \bar{Y}_{mR}\right)\left(Y_{nI} - \bar{Y}_{nI}\right)\right\} &= -E\left\{\left(Y_{nI} - \bar{Y}_{nI}\right)\left(Y_{mI} - \bar{Y}_{mI}\right)\right\}
\end{aligned}
\tag{9.34}
$$

Then, denoting the complex covariance matrix of **V** by $\mathbf{L} = E\left\{(\mathbf{V}-\overline{\mathbf{V}})(\mathbf{V}-\overline{\mathbf{V}})^{*T}\right\}$ assumed to be nonsingular, the CF of D is given by [17]

$$\Psi_D(\omega) = \left|\mathbf{I} - j\omega \mathbf{LQ}\right|^{-1} \exp\left\{-\overline{\mathbf{V}}^{*T}\mathbf{L}^{-1}\left[\mathbf{I} - (\mathbf{I} - j\omega \mathbf{LQ})^{-1}\right]\overline{\mathbf{V}}\right\} \qquad (9.35)$$

where **I** is the identity matrix.

The quadratic form in the argument of the exponential of (9.35) can be diagonalized by use of the transformation $\overline{\mathbf{V}} = \mathbf{U}_1\mathbf{N}^{-1}\mathbf{U}_2\mathbf{F}$ where $\mathbf{U}_1$ is the normalized modal matrix of $\mathbf{L}^{-1}$, $\mathbf{N}^2 = \mathbf{U}_1^T\mathbf{L}^{-1}\mathbf{U}_1$, and $\mathbf{U}_2$ is the normalized modal matrix of $\mathbf{N}^{-1}\mathbf{U}_1^{*T}\mathbf{Q}\mathbf{U}_1\mathbf{N}^{-1}$. Then, alternatively to (9.35)

$$\Psi_D(\omega) = \prod_{k=1}^{2n} \frac{\exp\left[\dfrac{j\omega\lambda_k|f_k|^2}{1 - j\omega\lambda_k}\right]}{1 - j\omega\lambda_k} \qquad (9.36)$$

where the f_k's are the elements of **F** and the λ_k's are the eigenvalues of **LQ**.

A special case of (9.36) corresponds to **X** and **Y** having zero mean in which case $\overline{\mathbf{V}} = 0$. For this case, the f_k's are all equal to zero and thus (9.36) simplifies to

$$\Psi_D(\omega) = \prod_{k=1}^{2n} \frac{1}{1 - j\omega\lambda_k} \qquad (9.37)$$

whose singularities consist solely of a finite number of finite-order poles.[6] Furthermore, the mean and variance of D are given by

$$E\{D\} = \sum_{k=1}^{2n} \lambda_k, \quad \mathrm{var}\, D = \sum_{k=1}^{2n} \lambda_k^2 \qquad (9.38)$$

In the most general case of nonzero mean **X** and **Y** vectors, determining the PDF of D from the inverse Fourier transform of $\Psi_D(\omega)$ is difficult because of the complicated nature of the exponential

---

[6] The appearance of poles in $\Psi_D(\omega)$ when $\overline{\mathbf{V}} = 0$ occurs because of the constraints placed on the real and imaginary parts of the components of **V** as in (9.33) and (9.34).

factor in (9.36). However, for the zero mean vector case, once the eigenvalues of **LQ** are found, $p_D(d)$ is readily determined from (9.37) through the use of the residue theorem. For example, if all eigenvalues have multiplicity one, then the partial fraction expansion of (9.37) is given by

$$\Psi_D(\omega) = \sum_{k=1}^{2n} \frac{C_k}{1 - j\omega\lambda_k}, \quad C_k = \prod_{\substack{i=1 \\ i \neq k}}^{2n} \frac{\lambda_k}{\lambda_k - \lambda_i} \tag{9.39}$$

with corresponding PDF

$$p_D(d) = \begin{cases} \sum_{k \in K_+} \frac{C_k}{\lambda_k} \exp\left(-\frac{d}{\lambda_k}\right), & d > 0 \\[2ex] -\sum_{k \in K_-} \frac{C_k}{\lambda_k} \exp\left(-\frac{d}{\lambda_k}\right), & d < 0 \end{cases} \tag{9.40}$$

where K_+ is the set of k corresponding to positive eigenvalues and K_- is the set of all k corresponding to negative eigenvalues. Furthermore,

$$E\{D^l\} = l! \sum_{k=1}^{2n} C_k \lambda_k^l \tag{9.41}$$

10

OTHER MISCELLANEOUS FORMS

A. Independent Rayleigh (+) Rayleigh

Let $R_1 = \|\mathbf{X}^{(1)}\|$ and $R_2 = \|\mathbf{X}^{(2)}\|$ be independent Rayleigh RVs corresponding to Gaussian vectors $\mathbf{X}^{(1)} \in N_2(0,\sigma^2)$ and $\mathbf{X}^{(2)} \in N_2(0,\sigma^2)$. Then, the sum RV $R = R_1 + R_2$ has the PDF

$$p_R(r) = \frac{1}{\sigma}\exp\left(-\frac{r^2}{4\sigma^2}\right)\left\{\sqrt{\pi}\left(\frac{r^2}{4\sigma^2}-\frac{1}{2}\right)\left[1-2Q\left(\frac{r}{\sqrt{2}\sigma}\right)\right]\right.$$
$$\left.+\frac{r}{2\sigma}\exp\left(-\frac{r^2}{4\sigma^2}\right)\right\}, \quad r\geq 0$$

(10.1)

B. Independent Gaussian (×) Rayleigh

Let $X \in N_1(0,\sigma_1^2)$ and $R=\|\mathbf{X}^{(2)}\|$ be independent Gaussian and Rayleigh RVs where $\mathbf{X}^{(2)} \in N_2(0,\sigma_2^2)$. Then, the product RV $Z = XR$ has the PDF

$$p_Z(z) = \frac{1}{2\sigma_1\sigma_2}\exp\left(-\frac{|z|}{\sigma_1\sigma_2}\right)$$

(10.2)

C. Independent Gaussian (×) Rayleigh (+) Gaussian

Let $X_1 \in N_1(0, \sigma_1^2)$, $R = \|\mathbf{X}^{(2)}\|$ and $X_3 \in N_1(0, \sigma_3^2)$ be independent Gaussian, Rayleigh and Gaussian RVs and $\mathbf{X}^{(2)} \in N_2(0, \sigma_2^2)$. Then, the RV $Z = X_1 R + X_3$ has the PDF

$$p_Z(z) = \frac{1}{2\sigma_1\sigma_2} \left\{ \exp\left[\frac{1}{2}\left(\frac{\sigma_3}{\sigma_1\sigma_2}\right)^2 - \frac{z}{\sigma_1\sigma_2} \right] Q\left(\frac{\sigma_3}{\sigma_1\sigma_2} - \frac{z}{\sigma_3}\right) \right.$$

$$\left. + \exp\left[\frac{1}{2}\left(\frac{\sigma_3}{\sigma_1\sigma_2}\right)^2 + \frac{z}{\sigma_1\sigma_2} \right] Q\left(\frac{\sigma_3}{\sigma_1\sigma_2} + \frac{z}{\sigma_3}\right) \right\}$$

(10.3)

D. Independent Gaussian (+) Rayleigh

Let $X \in N_1(0, \sigma_1^2)$ and $R = \|\mathbf{X}^{(2)}\|$ be independent Gaussian and Rayleigh RVs where $\mathbf{X}^{(2)} \in N_2(0, \sigma_2^2)$. Then, the RV $Z = X + R$ has the PDF

$$p_Z(z) = \frac{\sigma_1 \exp\left(-\frac{z^2}{2\sigma_1^2}\right)}{\sqrt{2\pi}(\sigma_1^2 + \sigma_2^2)} + \frac{\sigma_2 z \exp\left(-\frac{z^2\sigma_2^2}{2(\sigma_1^2 + \sigma_2^2)}\right)}{(\sigma_1^2 + \sigma_2^2)^{3/2}} \left[1 - Q\left(\frac{z\sigma_2}{\sigma_1\sqrt{\sigma_1^2 + \sigma_2^2}}\right)\right] \quad (10.4)$$

E. General Products of Ratios of Independent Gaussians

Let $X_k \in N_1(0, \sigma_x^2)$, $Y_k \in N_1(0, \sigma_y^2)$, $k = 1, 2, ..., n$ be mutually independent Gaussian RVs. Define the RV

$$Z = \prod_{k=1}^{n} \frac{X_k}{Y_k} \qquad (10.5)$$

Then, the PDF of Z is as follows:

1. $n = 1$

$$p_z(z) = \frac{\sigma_x \sigma_y}{\pi \left(\sigma_y^2 z^2 + \sigma_x^2 \right)} \tag{10.6}$$

2. $n = 2$

$$p_z(z) = \frac{\sigma_y^2}{\pi^2 \left(\sigma_y^2 z^2 - \sigma_x^2 \right)} \ln\left(\frac{\sigma_y^2 z^2}{\sigma_x^2} \right) \tag{10.7}$$

3. $n = 3$

$$p_z(z) = \frac{\sigma_y^2}{2!\pi^3 \left(\sigma_y^2 z^2 + \sigma_x^2 \right)} \left\{ \left[\ln\left(\frac{\sigma_y^2 z^2}{\sigma_x^2} \right) \right]^2 + 1 \right\} \tag{10.8}$$

4. $n = 4$

$$p_z(z) = \frac{\sigma_y^2}{3!\pi^4 \left(\sigma_y^2 z^2 - \sigma_x^2 \right)} \left\{ \left[\ln\left(\frac{\sigma_y^2 z^2}{\sigma_x^2} \right) \right]^3 + 4 \ln\left(\frac{\sigma_y^2 z^2}{\sigma_x^2} \right) \right\} \tag{10.9}$$

APPENDIX A: ALTERNATIVE FORMS

In this appendix, we present a number of alternative forms for the Gaussian Q-function, Marcum Q-function and incomplete gamma function. The motivation behind these forms comes from applications where it is desirable to have an integral representation in which the argument of the function is neither in the upper nor lower limit and furthermore appears in the integrand as an argument of an elementary function such as exponentials or trigonometrics.

1. One-Dimensional Distributions and Functions

A. The Gaussian Q-Function

The Gaussian Q-function, $Q(x)$, which is defined as the complement (with respect to unity) of the CDF corresponding to a Gaussian RV, has the canonical representation given in (1.3). In principle, this representation suffers from two disadvantages. First, from a computational standpoint, (1.3) requires truncation of the upper infinite limit when using numerical integral evaluation or algorithmic techniques. Second, the presence of the argument of the function as the lower limit of the integral poses analytical difficulties when this argument depends on other random parameters that ultimately require statistical averaging over their probability distributions. Such is the case, for example, when studying the performance of digital communication systems over additive white Gaussian noise (AWGN) channels that are also perturbed by random fading [15]. In such instances, what would clearly be more desirable would be to have a form for $Q(x)$ wherein the argument of the function is neither in the upper nor the lower limit of the integral and furthermore appears in the integrand as the argument of an elementary function. Still more desirable would be a form wherein the argument-independent limits are finite and the integrand maintains its Gaussian nature.

Although the existence of such a form dates back to a 1972 classified report (which has since become unclassified) by Nuttall [9][1], its popularity and recent extensive application primarily stem from its reintroduction by Craig [20] who presented it as a by-product of his work on the evaluation of the average probability of error performance for the two-dimensional AWGN channel. Specifically, the so-called Craig representation for $Q(x)$ (but only for $x \geq 0$) is given by

$$Q(x) = \frac{1}{\pi} \int_0^{\pi/2} \exp\left(-\frac{x^2}{2\sin^2\theta}\right) d\theta \qquad (A.1)$$

The form in (A.1) is not readily obtainable by a change of variables directly in (1.3). However, by first extending (1.3) to two dimensions (x and y) where one of the dimensions (y) is integrated over the half plane, a change of variables from rectangular to polar coordinates readily produces (A.1). Furthermore, (A.1) can be obtained directly by a straightforward change of variables of a standard known integral involving $Q(x)$, in particular, [1, Eq. (3.363.2)]. A description of both of these derivations can be found in [15, Appendix 4A]. Yet another derivation of (A.1) is given in Ref. 21 and is based on the fact that since the product of two independent random variables, one of which is a Rayleigh and the other a sinusoidal random process with random phase, is a Gaussian random variable, determining the CDF of this product variable is equivalent to evaluating the Gaussian Q-function.

B. The Marcum Q-Function

The first-order Marcum Q-function, $Q(\alpha, \beta)$, which is defined as the complement (with respect to unity) of the CDF corresponding to a normalized noncentral chi-square random variable with two degrees of freedom has the canonical representation given in (2.20). It is of interest to note that the complement (with respect to unity) of the

---

[1] The relation given there is actually for the complementary error function which is related to the Gaussian Q-function by (1.4). Other early traces of this alternative form can be found in the work of Weinstein [18] and Pawula [19].

first-order Marcum Q-function can be looked upon as a special case of the incomplete Toronto function [22, pp. 227–228] which finds its roots in the radar literature and is defined by

$$T_B(m,n,r) = 2r^{n-m+1}e^{-r^2}\int_0^B t^{m-n}e^{-t^2}I_n(2rt)dt \tag{A.2}$$

In particular, we have

$$T_{\frac{\beta}{\sqrt{2}}}\left(1,0,\tfrac{\alpha}{\sqrt{2}}\right) = 1 - Q_1(\alpha,\beta) \tag{A.3}$$

Furthermore, as $\beta \to \infty$, $Q_1(\alpha,\beta)$ can be related to the Gaussian Q-function as follows. Using the asymptotic (for large argument) form of the zero-order modified Bessel function of the first kind, we get [9, Eq. (A-27)]

$$Q_1(\alpha,\beta) \cong \int_\beta^\infty x\exp\left(-\frac{x^2+\alpha^2}{2}\right)\frac{\exp(\alpha x)}{\sqrt{2\pi\alpha x}}dx \cong \sqrt{\frac{\beta}{\alpha}}\frac{1}{\sqrt{2\pi}}\int_\beta^\infty \exp\left[-\frac{(x-\alpha)^2}{2}\right]dx$$

$$= \sqrt{\frac{\beta}{\alpha}}Q(\beta-\alpha) \tag{A.4}$$

Using integration by parts, it can also be shown that the first-order Marcum Q-function has the series form

$$Q_1(\alpha,\beta) = \exp\left(-\frac{\alpha^2+\beta^2}{2}\right)\sum_{k=0}^\infty \left(\frac{\alpha}{\beta}\right)^k I_k(\alpha\beta) \tag{A.5}$$

A comparison of (2.20) with (1.3) reveals that the canonical form of the Marcum Q-function suffers from the same two disadvantages as previously discussed for the Gaussian Q-function. Once again from the standpoint of application, it would be desirable to have an integral form of the Marcum Q-function in which the limits are finite and the integrand possesses a Gaussian nature. Historically, the roots of such an alternative form can be traced back to a 1965 report by Stein [23]; however, the more popular and available disclosures from which a recent flurry of applications have stemmed are found in Refs. 24, 25. Specifically, it has been shown that the first-order Marcum Q-function has the desirable representation

$$Q_1(\alpha, \beta) = \frac{1}{2\pi} \int_{-\pi}^{\pi} \left[\frac{1 + \left(\frac{\alpha}{\beta} \right) \sin \theta}{1 + 2 \left(\frac{\alpha}{\beta} \right) \sin \theta + \left(\frac{\alpha}{\beta} \right)^2} \right]$$

$$\times \exp\left[-\frac{\beta^2}{2} \left(1 + 2 \left(\frac{\alpha}{\beta} \right) \sin \theta + \left(\frac{\alpha}{\beta} \right)^2 \right) \right] d\theta, \ \beta > \alpha \geq 0$$

(A.6)

An analogous development to that used in arriving at (A.6) yields the result[2]

$$Q_1(\alpha, \beta) = 1 + \frac{1}{2\pi} \int_{-\pi}^{\pi} \left[\frac{\left(\frac{\beta}{\alpha} \right)^2 + \left(\frac{\beta}{\alpha} \right) \sin \theta}{1 + 2 \left(\frac{\beta}{\alpha} \right) \sin \theta + \left(\frac{\beta}{\alpha} \right)^2} \right]$$

$$\times \exp\left[-\frac{\alpha^2}{2} \left(1 + 2 \left(\frac{\beta}{\alpha} \right) \sin \theta + \left(\frac{\beta}{\alpha} \right)^2 \right) \right] d\theta, \ \alpha > \beta \geq 0$$

(A.7)

The results in (A.6) and (A.7) can put in a form with a more reduced integration interval. In particular, using the symmetry properties of the trigonometric functions over the intervals $(-\pi, 0)$ and $(0, \pi)$, we obtain the alternative forms

$$Q_1(\alpha, \beta) = \frac{1}{\pi} \int_0^{\pi} \left[\frac{1 \pm \left(\frac{\alpha}{\beta} \right) \cos \theta}{1 \pm 2 \left(\frac{\alpha}{\beta} \right) \cos \theta + \left(\frac{\alpha}{\beta} \right)^2} \right] \exp\left[-\frac{\beta^2}{2} \left(1 \pm 2 \left(\frac{\alpha}{\beta} \right) \cos \theta + \left(\frac{\alpha}{\beta} \right)^2 \right) \right] d\theta,$$

$$\beta > \alpha \geq 0$$

(A.8)

---

[2] At first glance it might appear from (A.7) that the Marcum-Q function can exceed unity. However, the integral in (A.7) is always less than or equal to zero. It should also be noted that the results in (A.6) and (A.7) can also be obtained from the work of Pawula [26] dealing with the relation between the Rice Ie-function and the Marcum Q-function.

and

$$Q_1(\alpha,\beta) = 1$$

$$+ \frac{1}{\pi} \int_0^\pi \frac{\left(\dfrac{\beta}{\alpha}\right)^2 \pm \left(\dfrac{\beta}{\alpha}\right)\cos\theta}{1 \pm 2\left(\dfrac{\beta}{\alpha}\right)\cos\theta + \left(\dfrac{\beta}{\alpha}\right)^2} \exp\left[-\frac{\alpha^2}{2}\left(1 \pm 2\left(\frac{\beta}{\alpha}\right)\cos\theta + \left(\frac{\beta}{\alpha}\right)^2\right)\right] d\theta, \text{(A.9)}$$

$$\alpha > \beta \geq 0$$

Before moving on to the generalized (mth order) Marcum Q-function, we discuss another alternative, and in some sense simpler, form of the first-order function that dispenses with the trigonometric factor that precedes the exponential in the integrands of (A.6) and (A.7) in favor of the sum of two purely exponential integrands each still having the desired dependence on α or β as appropriate. This relatively recent discovery due to Pawula [27] is given as

$$Q_1(\alpha,\beta) = \frac{1}{4\pi} \int_{-\pi}^\pi \left\{ \exp\left[-\frac{\beta^2}{2}\left(1 + 2\left(\frac{\alpha}{\beta}\right)\sin\theta + \left(\frac{\alpha}{\beta}\right)^2\right)\right] \right.$$

$$\left. + \exp\left[-\frac{\beta^2}{2}\left(1 - \left(\frac{\alpha}{\beta}\right)^2\right)^2\right]\left(1 + 2\left(\frac{\alpha}{\beta}\right)\sin\theta + \left(\frac{\alpha}{\beta}\right)^2\right)^{-1} \right\} d\theta, \text{(A.10)}$$

$$\beta \geq \alpha \geq 0$$

and

$$Q_1(\alpha,\beta) = 1 + \frac{1}{4\pi} \int_{-\pi}^\pi \left\{ \exp\left[-\frac{\alpha^2}{2}\left(1 + 2\left(\frac{\beta}{\alpha}\right)\sin\theta + \left(\frac{\beta}{\alpha}\right)^2\right)\right] \right.$$

$$\left. - \exp\left[-\frac{\alpha^2}{2}\left(1 - \left(\frac{\beta}{\alpha}\right)^2\right)^2\right]\left(1 + 2\left(\frac{\beta}{\alpha}\right)\sin\theta + \left(\frac{\beta}{\alpha}\right)^2\right)^{-1} \right\} d\theta, \text{(A.11)}$$

$$\alpha \geq \beta \geq 0$$

or equivalently in the reduced forms analogous to (A.8) and (A.9)

$$Q_1(\alpha,\beta) = \frac{1}{2\pi}\int_0^\pi \left\{ \exp\left[-\frac{\beta^2}{2}\left(1 \pm 2\left(\frac{\alpha}{\beta}\right)\cos\theta + \left(\frac{\alpha}{\beta}\right)^2 \right) \right] \right.$$

$$\left. + \exp\left[-\frac{\beta^2}{2}\left(1 - \left(\frac{\alpha}{\beta}\right)^2 \right) \right]^2 \left(1 \pm 2\left(\frac{\alpha}{\beta}\right)\cos\theta + \left(\frac{\alpha}{\beta}\right)^2 \right)^{-1} \right\} d\theta, \text{(A.12)}$$

$$\beta \ge \alpha \ge 0$$

and

$$Q_1(\alpha,\beta) = 1 + \frac{1}{2\pi}\int_0^\pi \left\{ \exp\left[-\frac{\alpha^2}{2}\left(1 \pm 2\left(\frac{\beta}{\alpha}\right)\cos\theta + \left(\frac{\beta}{\alpha}\right)^2 \right) \right] \right.$$

$$\left. - \exp\left[-\frac{\alpha^2}{2}\left(1 - \left(\frac{\beta}{\alpha}\right)^2 \right) \right]^2 \left(1 \pm 2\left(\frac{\beta}{\alpha}\right)\cos\theta + \left(\frac{\beta}{\alpha}\right)^2 \right)^{-1} \right\} d\theta, \text{(A.13)}$$

$$\alpha \ge \beta \ge 0$$

Since the first exponential integrand in each of (A.10) through (A.13) is identical to the exponential integrand in the corresponding equations (A.6) through (A.9), we can look upon the second exponential in the integrands of the former group of equations as compensating for the lack of the trigonometric multiplying factor in the integrands of the latter equation group.

The generalized (mth-order) Marcum Q-function, $Q_m(\alpha,\beta)$, which is defined as the complement (with respect to unity) of the CDF corresponding to a normalized noncentral chi-square random variable with $m+1$ degrees of freedom has the canonical representation given in (2.24). Once again, it is of interest to note that the complement (with respect to unity) of the generalized Marcum Q-function can be looked upon as a special case of the incomplete Toronto function. Specifically, analogous to (A.3),

$$T_{\frac{\beta}{\sqrt{2}}}\left(2m-1, m-1, \tfrac{\alpha}{\sqrt{2}}\right) = 1 - Q_m(\alpha,\beta) \tag{A.14}$$

Furthermore, as $\beta \to \infty$, $Q_m(\alpha,\beta)$ can be related to the generalized Gaussian Q-function in the same manner as was done for the first-order Marcum Q-function. Specifically, since the asymptotic (for large argument) form of the kth-order modified Bessel function of the first kind is independent of the order, then

$$Q_m(\alpha,\beta) \cong \int_\beta^\infty x\left(\frac{x}{\alpha}\right)^{m-1} \exp\left(-\frac{x^2+\alpha^2}{2}\right)\frac{\exp(\alpha x)}{\sqrt{2\pi\alpha x}}dx$$

$$\cong \left(\frac{\beta}{\alpha}\right)^{m-\frac{1}{2}}\frac{1}{\sqrt{2\pi}}\int_\beta^\infty \exp\left[-\frac{(x-\alpha)^2}{2}\right]dx = \left(\frac{\beta}{\alpha}\right)^{m-1/2}Q(\beta-\alpha)$$

(A.15)

A comparison of (2.24) with (1.3) and (2.20) reveals that the canonical form of the generalized Marcum Q-function suffers from the same two disadvantages as previously discussed for the Gaussian Q-function and the first-order Marcum Q-function. Once again from the standpoint of application, it would be desirable to have an integral form for the generalized Marcum Q-function in which the limits are finite and the integrand possesses a Gaussian nature. The discovery of such a form was made independently in Refs. 24 and 25 with the following results:

$$Q_m(\alpha,\beta) = \frac{1}{2\pi}\int_{-\pi}^{\pi}\frac{\left(\frac{\alpha}{\beta}\right)^{-(m-1)}\left\{\cos[(m-1)(\theta+\frac{\pi}{2})]-\left(\frac{\alpha}{\beta}\right)\cos[m(\theta+\frac{\pi}{2})]\right\}}{1+2\left(\frac{\alpha}{\beta}\right)\sin\theta+\left(\frac{\alpha}{\beta}\right)^2}$$

$$\times\exp\left[-\frac{\beta^2}{2}\left(1+2\left(\frac{\alpha}{\beta}\right)\sin\theta+\left(\frac{\alpha}{\beta}\right)^2\right)\right]d\theta, \quad 0^+\le\alpha/\beta<1$$

(A.16)

and

$$Q_m(\alpha,\beta) = 1-\frac{1}{2\pi}\int_{-\pi}^{\pi}\frac{\left(\frac{\beta}{\alpha}\right)^m\left\{\cos[m(\theta+\frac{\pi}{2})]-\left(\frac{\beta}{\alpha}\right)\cos[(m-1)(\theta+\frac{\pi}{2})]\right\}}{1+2\left(\frac{\beta}{\alpha}\right)\sin\theta+\left(\frac{\beta}{\alpha}\right)^2}$$

$$\times\exp\left[-\frac{\alpha^2}{2}\left(1+2\left(\frac{\beta}{\alpha}\right)\sin\theta+\left(\frac{\beta}{\alpha}\right)^2\right)\right]d\theta, \quad 0\le\beta/\alpha<1$$

(A.17)

that can be further simplified and separated into m odd and m even as

$$Q_m(\alpha,\beta) = \frac{1}{2\pi}\int_{-\pi}^{\pi} \frac{(-1)^{\frac{m-1}{2}}\left(\dfrac{\alpha}{\beta}\right)^{-(m-1)}\left[\cos(m-1)\theta + \left(\dfrac{\alpha}{\beta}\right)\sin m\theta\right]}{1+2\left(\dfrac{\alpha}{\beta}\right)\sin\theta + \left(\dfrac{\alpha}{\beta}\right)^2}$$

$$\times \exp\left[-\frac{\beta^2}{2}\left(1+2\left(\frac{\alpha}{\beta}\right)\sin\theta + \left(\frac{\alpha}{\beta}\right)^2\right)\right] d\theta,\ 0^+ < \alpha/\beta < 1,\ m\ \text{odd}$$

$$(A.18)$$

$$Q_m(\alpha,\beta) = \frac{1}{2\pi}\int_{-\pi}^{\pi} \frac{(-1)^{\frac{m}{2}}\left(\dfrac{\alpha}{\beta}\right)^{-(m-1)}\left[\sin(m-1)\theta - \left(\dfrac{\alpha}{\beta}\right)\cos m\theta\right]}{1+2\left(\dfrac{\alpha}{\beta}\right)\sin\theta + \left(\dfrac{\alpha}{\beta}\right)^2}$$

$$\times \exp\left[-\frac{\beta^2}{2}\left(1+2\left(\frac{\alpha}{\beta}\right)\sin\theta + \left(\frac{\alpha}{\beta}\right)^2\right)\right] d\theta,\ 0^+ < \alpha/\beta < 1,\ m\ \text{even}$$

$$(A.19)$$

and

$$Q_m(\alpha,\beta) = 1 + \frac{1}{2\pi}\int_{-\pi}^{\pi} \frac{(-1)^{\frac{m-1}{2}}\left(\dfrac{\beta}{\alpha}\right)^{m}\left[\sin m\theta + \left(\dfrac{\beta}{\alpha}\right)\cos(m-1)\theta\right]}{1+2\left(\dfrac{\beta}{\alpha}\right)\sin\theta + \left(\dfrac{\beta}{\alpha}\right)^2}$$

$$(A.20)$$

$$\times \exp\left[-\frac{\alpha^2}{2}\left(1+2\left(\frac{\beta}{\alpha}\right)\sin\theta + \left(\frac{\beta}{\alpha}\right)^2\right)\right] d\theta,\ 0 \le \beta/\alpha < 1,\ m\ \text{odd}$$

$$Q_m(\alpha,\beta) = 1 - \frac{1}{2\pi}\int_{-\pi}^{\pi} \frac{(-1)^{\frac{m}{2}}\left(\dfrac{\beta}{\alpha}\right)^{m}\left[\cos m\theta - \left(\dfrac{\beta}{\alpha}\right)\sin(m-1)\theta\right]}{1+2\left(\dfrac{\beta}{\alpha}\right)\sin\theta + \left(\dfrac{\beta}{\alpha}\right)^2}$$

$$(A.21)$$

$$\times \exp\left[-\frac{\alpha^2}{2}\left(1+2\left(\frac{\beta}{\alpha}\right)\sin\theta + \left(\frac{\beta}{\alpha}\right)^2\right)\right] d\theta,\ 0 \le \beta/\alpha < 1,\ m\ \text{even}$$

We observe from (A.16) through (A.21) that α and β are restricted to be unequal. The special case of $\alpha = \beta$ has the closed-form result [24]

$$Q_m(\alpha,\alpha) = \frac{1}{2} + \exp(-\alpha^2)\left[\frac{I_0(\alpha^2)}{2} + \sum_{k=1}^{m-1} I_k(\alpha^2)\right] \qquad (A.22)$$

Finally, we note that unlike the first-order Marcum Q-function, additional simplification as in (A.8) and (A.9) are not readily possible for the mth-order function.

C. The Nuttall Q-Function

The Nuttall Q-function [9] is a generalization of the Marcum Q-function and is defined in (4.41) together with its recursive relation in (4.42). Since the Nuttall Q-function is not considered to be a tabulated function, it is desirable to express the recursion relation of (4.42) in a form that allows solution in terms of the first-order Marcum Q-function which has been tabulated [8] or equivalently in terms of the generalized Marcum Q-function. The possibility of doing such when $m+n$ is odd was suggested in Ref. 9; however, the explicit solution was not provided. Letting $m = n+2k+1$ in which case $m+n = 2(n+k)+1$ and is therefore odd, an explicit expression for $Q_{n+2k+1,n}(\alpha,\beta)$ in terms of only the first-order Marcum Q-function and modified Bessel functions of the first kind was found by this author [28]. Specifically,

$$Q_{n+2k+1,n}(\alpha,\beta) = 2^k k! \alpha^n L_k^{(n)}(-\alpha^2/2)\left[\overbrace{Q(\alpha,\beta) + \exp\left(-\frac{\alpha^2+\beta^2}{2}\right)\sum_{i=1}^{n}\left(\frac{\alpha}{\beta}\right)^i I_i(\alpha\beta)}^{Q_{n+1}(\alpha,\beta)}\right]$$

$$+ \exp\left(-\frac{\alpha^2+\beta^2}{2}\right)\beta^n\left[2^k k! \sum_{l=1}^{k}\left(L_l^{(n)}(-\alpha^2/2) - L_{l-1}^{(n)}(-\alpha^2/2)\right)\right.$$

$$\left.\times\left(\frac{\beta}{\alpha}\right)^l I_{n+l}(\alpha\beta) + \sum_{l=1}^{k} P_{k,l}(\beta^2)(\alpha\beta)^l\left(\frac{\beta}{\alpha}\right)^l I_{n+l-1}(\alpha\beta)\right]$$

$$(A.23)$$

where

$$L_k^{(n)}(x) \triangleq \sum_{i=0}^{k} \binom{k+n}{k-i} \frac{(-x)^i}{i!} \tag{A.24}$$

is the generalized (kth –order) Laguerre polynomial [1] and $P_{k,l}(\beta^2)$ is a polynomial of order $k-l$ in β^2, namely,

$$P_{k,l}(\beta^2) = \sum_{j=0}^{k-l} d_j(k,l)\beta^{2j} \tag{A.25}$$

with

$$d_j(k,l) \triangleq 2^{k-l-j} \binom{k-1-j}{k-l-j} \frac{(n+k)!}{(n+l+j)!} = 2^{k-l-j} \frac{(k-1-j)!}{(l-1)!} \binom{k+n}{k-l-j} \tag{A.26}$$

Note that for $k=0$, $L_0^{(n)}(x)=1$ and (A.23) reduces, as it should, to (4.43). Also, the two weighted sums of Bessel functions in (A.23) could, in principle, be combined into a single sum of weighted Bessel functions; however, the additional notational complexity needed to do this is probably not worth it. Finally, since $\exp[-(a^2+b^2)/2]I_n(ab)$ can be expressed in the form

$$\exp\left(-\frac{\alpha^2+\beta^2}{2}\right) I_n(\alpha\beta) = \frac{1}{2\pi} \int_{-\pi}^{\pi} \cos[n(\theta+\tfrac{\pi}{2})]$$

$$\times \exp\left\{-\frac{\beta^2}{2}\left[1+2\left(\frac{\alpha}{\beta}\right)\sin\theta+\left(\frac{\alpha}{\beta}\right)^2\right]\right\} d\theta \tag{A.27}$$

then $Q_{n+2k+1,n}(\alpha,\beta)$ can be put into the form of say (A.16), namely,

$$Q_{n+2k+1,n}(\alpha,\beta) = \frac{1}{2\pi} \int_{-\pi}^{\pi} g_m(\theta;\alpha,\beta)$$

$$\times \exp\left\{-\frac{\beta^2}{2}\left[1+2\left(\frac{a}{\beta}\right)\alpha\sin\theta+\left(\frac{\alpha}{\beta}\right)^2\right]\right\} d\theta \tag{A.28}$$

where $g_m(\theta;\alpha,\beta)$ is a purely trigonometric function of θ.

Unfortunately, for $m+n$ even, as pointed out by Nuttall [9], a solution for $Q_{m,n}(a,b)$, e.g., $Q_{n+2k,n}(a,b)$ with k an arbitrary integer, requires having two fundamental functions, either $Q_{0,0}(a,b)$ and $Q_{2,0}(a,b)$ or $Q_{0,0}(a,b)$ and $Q_{1,1}(a,b)$, neither pair of which is related to the Marcum Q-function or other tabulated functions. Thus, it is not possible to find a solution for $Q_{m,n}(a,b)$ in the form of (A.28).

D. The Complementary Incomplete Gamma Function

The complementary incomplete gamma function, $\Gamma(\alpha,x)$, is defined by

$$\Gamma(\alpha,x) = \int_x^\infty t^{\alpha-1} \exp(-t)dt \tag{A.29}$$

or equivalently, letting $t = y^2/2$,

$$\Gamma(\alpha,x) = \frac{1}{2^{\alpha-1}} \int_{\sqrt{2x}}^\infty y^{2\alpha-1} \exp\left(-\frac{y^2}{2}\right)dy \tag{A.30}$$

Two special cases of this function are of interest. First for $\alpha = 1$, a normalized version of (A.29) gives the exponential function, i.e.,

$$\frac{\Gamma(1,x)}{2\Gamma(1)} = \frac{1}{2}\int_x^\infty \exp(-t)dt = \frac{1}{2}\exp(-x) \tag{A.31}$$

Second for $\alpha = 1/2$, a normalized version of (A.30) gives the Gaussian Q-function, i.e.,

$$\frac{\Gamma(1/2,x)}{\Gamma(1/2)} = \frac{1}{\sqrt{2\pi}} \int_{\sqrt{2x}}^\infty y^{2\alpha-1} \exp\left(-\frac{y^2}{2}\right)dy = Q\left(\sqrt{2x}\right) \tag{A.32}$$

Analogous to what was done for the Gaussian Q-function and Marcum Q-function, the incomplete gamma function can be put into a more desirable form where the limits of the integral are finite and the integrand retains its Gaussian nature. Specifically, it can be shown that $\Gamma(\alpha,x)$ has the form

$$\Gamma(\alpha, x) = 2\int_0^{\pi/2} \frac{\cos\theta}{(\sin\theta)^{1+2\alpha}} x^\alpha \exp\left(-\frac{x}{\sin^2\theta}\right) d\theta \qquad \text{(A.33)}$$

or equivalently

$$\frac{\Gamma(\alpha, x)}{2\Gamma(\alpha)} = \frac{1}{\Gamma(\alpha)}\int_0^{\pi/2} \frac{\cos\theta}{(\sin\theta)^{1+2\alpha}} x^\alpha \exp\left(-\frac{x}{\sin^2\theta}\right) d\theta \qquad \text{(A.34)}$$

Since from (A.1) we have

$$Q(\sqrt{2x}) = \frac{1}{\pi}\int_0^{\pi/2} \exp\left(-\frac{x}{\sin^2\theta}\right) d\theta \qquad \text{(A.35)}$$

then equating (A.35) with (A.32) and using the form in (A.34) gives yet another alternative form for the Gaussian Q-function, namely,

$$Q(x) = \frac{\Gamma(1/2, x^2/2)}{2\Gamma(1/2)} = \frac{1}{\sqrt{\pi}}\int_0^{\pi/2} \frac{\cos\theta}{(\sin\theta)^2} \frac{x}{\sqrt{2}} \exp\left(-\frac{x^2}{2\sin^2\theta}\right) d\theta \quad \text{(A.36)}$$

2. Two-Dimensional Distributions and Functions

A. The Gaussian Q-Function

The two-dimensional Gaussian Q-function, $Q(x, y; \rho)$, is defined by

$$Q(x, y; \rho) = \frac{1}{2\pi\sqrt{1-\rho^2}}\int_x^\infty \int_y^\infty \exp\left[-\frac{u^2 + v^2 - 2\rho uv}{2(1-\rho^2)}\right] dv\, du \qquad \text{(A.37)}$$

and is the complement of the two-dimensional CDF corresponding to the PDF in (3.2) with proper change of notation. Using a geometric approach similar to that used by Craig [20] in arriving at the alternative form of the one-dimensional Gaussian Q-function in (A.1), the author first derived an analogous form for $Q(x, y; \rho)$ in the region $x \geq 0, y \geq 0$ as [15, Eq. (4.7)]

$$Q(x,y;\rho) = \frac{1}{2\pi} \int_0^{\frac{\pi}{2}-\tan^{-1}y/x} \frac{\sqrt{1-\rho^2}}{1-\rho\sin 2\theta} \exp\left\{ -\frac{x^2}{2}\left(\frac{1}{1-\rho^2}\right)\frac{1-\rho\sin 2\theta}{\sin^2\theta} \right\} d\theta$$

$$+ \frac{1}{2\pi} \int_0^{\tan^{-1}y/x} \frac{\sqrt{1-\rho^2}}{1-\rho\sin 2\theta} \exp\left\{ -\frac{y^2}{2}\left(\frac{1}{1-\rho^2}\right)\frac{1-\rho\sin 2\theta}{\sin^2\theta} \right\} d\theta,$$

(A.38)

More recently, this author derived [28] another alternative form for $Q(x,y;\rho)$ in this same region that dispensed with the trigonometric factor that precedes the exponentials in the integrands of (A.38) and furthermore resulted in an exponential argument that is precisely in the same simple form as that in the Craig representation of $Q(x)$. Specifically,[3]

$$Q(x,y;\rho) = \frac{1}{2\pi} \int_0^{\tan^{-1}\left(\frac{\sqrt{1-\rho^2}x/y}{1-\rho x/y}\right)} \exp\left(-\frac{x^2}{2\sin^2\theta} \right) d\theta$$

$$+ \frac{1}{2\pi} \int_0^{\tan^{-1}\left(\frac{\sqrt{1-\rho^2}y/x}{1-\rho y/x}\right)} \exp\left(-\frac{y^2}{2\sin^2\theta} \right) d\theta, \quad x \geq 0, y \geq 0$$

(A.39)

Note that for $\rho \neq 0$ but $y = x$, (A.39) simplifies to

$$Q(x,x;\rho) = \frac{1}{\pi} \int_0^{\tan^{-1}\left(\sqrt{\frac{1+\rho}{1-\rho}}\right)} \exp\left(-\frac{x^2}{2\sin^2\theta} \right) d\theta$$

(A.40)

Also, for $\rho = 0$, we immediately get from (A.40) the Craig form for $Q^2(x)$, namely [25, Eq. (80)]

$$Q^2(x) = \frac{1}{\pi} \int_0^{\pi/4} \exp\left(-\frac{x^2}{2\sin^2\theta} \right) d\theta$$

(A.41)

whereas for $\rho = 1$, (A.40) becomes the Craig form for $Q(x)$.

To arrive at alternative forms analogous to (A.39) for other regions of x and y, we first note the following relations:

---

[3] Since $1-\rho x/y$ and $1-\rho y/x$ can take on positive or negative values, the arctangents in the upper limits of the integrals in (A.39) are defined by $\tan^{-1} X/Y = \pi(1 - \operatorname{sgn} Y)/2 + (\operatorname{sgn} Y)\tan^{-1} X/|Y|$.

$$Q(x,y;\rho) = Q(x) - Q(x,-y;-\rho), \; x \geq 0, y < 0 \tag{A.42}$$

$$Q(x,y;\rho) = Q(y) - Q(-x,y;-\rho), \; x < 0, y \geq 0 \tag{A.43}$$

$$Q(x,y;\rho) = 1 - Q(-x) - Q(-y) + Q(-x,-y;\rho), \; x < 0, y < 0 \tag{A.44}$$

Recognizing that because of the symmetry of the sinusoidal function, the alternative form of the one-dimensional Gaussian Q-function in (A.1) can also be written as

$$Q(x) = \frac{1}{2\pi} \int_0^\pi \exp\left(-\frac{x^2}{2\sin^2\theta}\right) d\theta, \; x \geq 0 \tag{A.45}$$

then substituting (A.45) and (A.39) in (A.42) – (A.44), we arrive at the following:

$$Q(x,y;\rho) = \frac{1}{2\pi} \int_{\tan^{-1}\left(-\frac{\sqrt{1-\rho^2}x/y}{1-\rho x/y}\right)}^\pi \exp\left(-\frac{x^2}{2\sin^2\theta}\right) d\theta$$

$$-\frac{1}{2\pi} \int_0^{\tan^{-1}\left(-\frac{\sqrt{1-\rho^2}y/x}{1-\rho y/x}\right)} \exp\left(-\frac{y^2}{2\sin^2\theta}\right) d\theta, \; x \geq 0, y < 0 \tag{A.46}$$

$$Q(x,y;\rho) = -\frac{1}{2\pi} \int_0^{\tan^{-1}\left(-\frac{\sqrt{1-\rho^2}x/y}{1-\rho x/y}\right)} \exp\left(-\frac{x^2}{2\sin^2\theta}\right) d\theta$$

$$+\frac{1}{2\pi} \int_{\tan^{-1}\left(-\frac{\sqrt{1-\rho^2}y/x}{1-\rho y/x}\right)}^\pi \exp\left(-\frac{y^2}{2\sin^2\theta}\right) d\theta, \; x < 0, y \geq 0 \tag{A.47}$$

$$Q(x,y;\rho) = 1 - \frac{1}{2\pi} \int_{\tan^{-1}\left(\frac{\sqrt{1-\rho^2}x/y}{1-\rho x/y}\right)}^\pi \exp\left(-\frac{x^2}{2\sin^2\theta}\right) d\theta$$

$$-\frac{1}{2\pi} \int_{\tan^{-1}\left(\frac{\sqrt{1-\rho^2}y/x}{1-\rho y/x}\right)}^\pi \exp\left(-\frac{y^2}{2\sin^2\theta}\right) d\theta, \; x < 0, y < 0 \tag{A.48}$$

APPENDIX B: INTEGRALS INVOLVING THE GAUSSIAN Q-FUNCTION AND THE MARCUM Q-FUNCTION

The material in this appendix is largely taken from a number of public documents (including the author's) [9, 15, 29, 30] some of which again, because of their age and origin or specialization of subject, may not be readily available or familiar to the average reader. It is for this reason that this appendix is being included in the book. Not all integrals involving these functions are included in the tabulation. Rather, emphasis is placed on those that less commonly appear in the published literature but are of interest in practical applications

1. The Gaussian Q-Function

Using the alternative form of the Gaussian Q-function as given in (A.1), we can express integrals of the form $\int_0^\infty Q(a\sqrt{y})g(y)dy$ in terms of the Laplace transform of $g(y)$, namely, $G(s) = \int_0^\infty e^{-sy}g(y)dy$, by[4]

$$\int_0^\infty g(y)Q(a\sqrt{y})dy = \frac{1}{\pi}\int_0^{\pi/2} G\left(\frac{a^2}{2\sin^2\theta}\right)d\theta, \quad a \geq 0 \qquad (B.1)$$

or equivalently

$$\int_0^\infty xg(x^2)Q(ax)dx = \frac{1}{2\pi}\int_0^{\pi/2} G\left(\frac{a^2}{2\sin^2\theta}\right)d\theta, \quad a \geq 0 \qquad (B.2)$$

---

[4] Quite often one is interested in integrals of this type where $g(y)$ takes the form of a PDF.

The advantage of (B.1) or (B.2) is that the integration now has finite limits and thus even if the Laplace transform is such that the integral cannot be obtained in closed form, it can still be evaluated quite simply by numerical integration. In this section of the appendix, we focus on functions $g(x)$ whose Laplace transform is known in closed form and as such the integral can be expressed either in the finite integral form of (B.1) or (B.2) or in closed form.

A. Q-Function and x

$$\int_0^\infty x^{2q-1} Q(ax)\,dx = \frac{2^{q-2}\,\Gamma(q+1/2)}{a^{2q}q\sqrt{\pi}}, a>0, q>0 \tag{B.3}$$

B. Q-Function with Exponentials and x

$$\int_0^\infty x \exp\left(-\frac{b^2 x^2}{2}\right) Q(ax)\,dx = \frac{1}{2b^2}\left[1-\sqrt{\frac{a^2}{a^2+b^2}}\right], a\geq 0 \tag{B.4}$$

$$\int_0^\infty x^{2m-1} \exp\left(-\frac{mb^2 x^2}{2}\right) Q(ax)\,dx = \frac{2^{m-2}}{(mb^2)^m}\left[1-\sqrt{\frac{a^2}{a^2+mb^2}}\right.$$

$$\left.\times \sum_{k=0}^{m-1}\binom{2k}{k}\left(\frac{mb^2}{4(a^2+mb^2)}\right)^k\right], a\geq 0, m \text{ integer} \tag{B.5}$$

$$\int_0^\infty x^{2m-1} \exp\left(-\frac{mb^2 x^2}{2}\right) Q(ax)\,dx = \frac{2^{m-2}\,\Gamma(m+1/2)}{m\sqrt{\pi}}\,\frac{a}{(a^2+mb^2)^{m+1/2}}$$

$$\times {}_2F_1\left(1, m+\frac{1}{2}; m+1; \frac{mb^2}{a^2+mb^2}\right), \tag{B.6}$$

$$a\geq 0, m \text{ noninteger}$$

$$\int_{-\infty}^\infty \exp\left(-a^2(x+\mu)^2+bx\right)Q(cx)\,dx = \frac{\sqrt{\pi}}{a}\exp\left(\frac{b^2}{4a^2}-b\mu\right)Q\left(\frac{c(b-2a^2\mu)}{\sqrt{2}a\sqrt{c^2+2a^2}}\right) \tag{B.7}$$

C. Q-Function with Exponentials, Bessel Functions and x

$$\int_0^\infty x \exp\left(-\frac{b^2 x^2}{2}\right) I_0(cx^2) Q(ax) dx =$$

$$\frac{1}{\pi} \int_0^{\pi/2} \left(\frac{\sin^4 \theta}{a^4 + 2a^2 b^2 \sin^2 \theta + b^2 (b^2 - 4c^2) \sin^4 \theta}\right)^{1/2} d\theta \tag{B.8}$$

$$\int_0^\infty x \exp\left(-\frac{b^2 x^2}{2}\right) I_0(cx) Q(ax) dx = \frac{1}{\pi} \int_0^{\pi/2} \left(\frac{\sin^2 \theta}{b^2 \sin^2 \theta + a^2}\right)^{1/2} \tag{B.9}$$

$$\times \exp\left[\frac{c^2}{2}\left(\frac{\sin^2 \theta}{b^2 \sin^2 \theta + a^2}\right)\right] d\theta, \ a \ge 0$$

2. The First-Order Marcum Q-Function

A. Q-Function with One Linear Argument

$$\int_0^\infty Q_1(b, ax) dx = \frac{\sqrt{2\pi}}{4a} b^2 \exp\left(-\frac{b^2}{4}\right)\left[\left(1 + \frac{2}{b^2}\right) I_0\left(\frac{b^2}{4}\right) + I_1\left(\frac{b^2}{4}\right)\right] \tag{B.10}$$

$$\int_0^\infty \left[1 - Q_1(ax, b)\right] dx = \frac{\sqrt{2\pi}}{4a} b^2 \exp\left(-\frac{b^2}{4}\right)\left[I_0\left(\frac{b^2}{4}\right) + I_1\left(\frac{b^2}{4}\right)\right] \tag{B.11}$$

B. Q-Function with One Linear Argument and Exponentials

$$\int_0^\infty \exp(-p^2 x^2 / 2) Q_1(b, ax) dx =$$

$$\sqrt{\frac{\pi}{2}} \frac{1}{p} \left[1 - 2Q_1\left(\frac{b}{2}\left(1 - \frac{p}{\sqrt{a^2 + p^2}} \right), \frac{b}{2}\left(1 + \frac{p}{\sqrt{a^2 + p^2}} \right) \right) \right.$$

$$\left. + \left(1 + \frac{p}{\sqrt{a^2 + p^2}} \right) \exp\left(-\frac{b^2}{4} \frac{a^2 + 2p^2}{a^2 + p^2} \right) I_0\left(\frac{a^2 b^2}{4(a^2 + p^2)} \right) \right]$$

$$(\text{B.12})$$

$$\int_0^\infty \exp(-p^2 x^2 / 2) Q_1(ax, b) dx = \sqrt{\frac{\pi}{2}} \frac{1}{p} \left[2Q_1\left(\frac{b}{2}\left(1 - \frac{p}{\sqrt{a^2 + p^2}} \right), \frac{b}{2}\left(1 + \frac{p}{\sqrt{a^2 + p^2}} \right) \right) \right.$$

$$\left. - \exp\left(-\frac{b^2}{4} \frac{a^2 + 2p^2}{a^2 + p^2} \right) I_0\left(\frac{a^2 b^2}{4(a^2 + p^2)} \right) \right]$$

$$(\text{B.13})$$

C. Q-Function with One Linear Argument and x

$$\int_c^\infty x Q_1(b, ax) dx = \frac{2 + b^2 - a^2 c^2}{2a^2} Q_1(b, ac) + \frac{c}{2a} \exp\left(-\frac{a^2 c^2 + b^2}{2} \right)$$
$$\times \left[ac I_0(abc) + b I_1(abc) \right] \qquad (\text{B.14})$$

$$\int_0^c x Q_1(b, ax) dx = \frac{2 + b^2}{2a^2} - \frac{2 + b^2 - a^2 c^2}{2a^2} Q_1(b, ac) - \frac{c}{2a} \exp\left(-\frac{a^2 c^2 + b^2}{2} \right)$$
$$\times \left[ac I_0(abc) + b I_1(abc) \right] \qquad (\text{B.15})$$

$$\int_0^c x Q_1(ax, b) dx = \frac{a^2 c^2 - b^2}{2a^2} \left[1 - Q_1(b, ac) \right] + \frac{c}{2a} \exp\left(-\frac{a^2 c^2 + b^2}{2} \right)$$
$$\times \left[ac I_0(abc) + b I_1(abc) \right] \qquad (\text{B.16})$$

$$\int_c^\infty x \left[1 - Q_1(ax, b) \right] dx = \frac{b^2 - a^2 c^2}{2a^2} Q_1(b, ac) + \frac{c}{2a} \exp\left(-\frac{a^2 c^2 + b^2}{2} \right)$$
$$\times \left[ac I_0(abc) + b I_1(abc) \right] \qquad (\text{B.17})$$

D. Q-Function with One Linear Argument, Exponentials and Powers of x

$$\int_c^\infty x\exp(-p^2x^2/2)Q_1(ax,b)dx = \frac{1}{p^2}\exp\left(-\frac{p^2c^2}{2}\right)Q_1(ac,b)$$

$$+\exp\left(-\frac{p^2b^2}{2(a^2+p^2)}\right)\left[1-Q_1\left(c\sqrt{a^2+p^2},\frac{ab}{\sqrt{a^2+p^2}}\right)\right]\Bigg\}$$
(B.18)

$$\int_0^c x\exp(-p^2x^2/2)Q_1(ax,b)dx =$$

$$\frac{1}{p^2}\left[\exp\left(-\frac{p^2b^2}{2(a^2+p^2)}\right)Q_1\left(c\sqrt{a^2+p^2},\frac{ab}{\sqrt{a^2+p^2}}\right)\right.$$
(B.19)

$$\left.-\exp\left(-\frac{p^2c^2}{2}\right)Q_1(ac,b)\right]$$

$$\int_0^c x\exp(p^2x^2/2)Q_1(ax,b)dx =$$

$$\frac{1}{p^2}\exp\left(\frac{p^2c^2}{2}\right)Q_1(ac,b)$$

$$+\exp\left(\frac{p^2b^2}{2(a^2-p^2)}\right)\left[1-Q_1\left(c\sqrt{a^2-p^2},\frac{ab}{\sqrt{a^2-p^2}}\right)\right]\Bigg\}, p\neq a$$
(B.20)

$$\int_0^c x\exp(p^2x^2/2)Q_1(px,b)dx = \frac{1}{p^2}\exp\left(\frac{p^2c^2}{2}\right)[1-Q_1(b,pc)]$$
(B.21)

$$\int_c^\infty x\exp(-p^2x^2/2)Q_1(b,ax)dx =$$

$$\frac{1}{p^2}\left[\exp\left(-\frac{p^2c^2}{2}\right)Q_1(b,ac)\right.$$
(B.22)

$$\left.-\frac{a^2}{a^2+p^2}\exp\left(-\frac{p^2b^2}{2(a^2+p^2)}\right)Q_1\left(\frac{ab}{\sqrt{a^2+p^2}},c\sqrt{a^2+p^2}\right)\right]$$

$$\int_0^c x \exp(-p^2 x^2 / 2) Q_1(b, ax) dx =$$

$$\frac{1}{p^2} \left\{ 1 - \exp\left(-\frac{p^2 c^2}{2} \right) Q_1(b, ac) \right.$$

$$\left. - \frac{a^2}{a^2 + p^2} \exp\left(-\frac{p^2 b^2}{2(a^2 + p^2)} \right) \left[1 - Q_1\left(\frac{ab}{\sqrt{a^2 + p^2}}, c\sqrt{a^2 + p^2} \right) \right] \right\}$$

$$(B.23)$$

$$\int_c^\infty x \exp(p^2 x^2 / 2) Q_1(b, ax) dx =$$

$$\frac{1}{p^2} \left[\frac{a^2}{a^2 - p^2} \exp\left(\frac{p^2 b^2}{2(a^2 - p^2)} \right) Q_1\left(\frac{ab}{\sqrt{a^2 - p^2}}, c\sqrt{a^2 - p^2} \right) \right. \quad (B.24)$$

$$\left. - \exp\left(\frac{p^2 c^2}{2} \right) Q_1(b, ac) \right], \, p < a$$

$$\int_0^c x \exp(p^2 x^2 / 2) Q_1(b, ax) dx =$$

$$\frac{1}{p^2} \left\{ \frac{a^2}{a^2 - p^2} \exp\left(\frac{p^2 b^2}{2(a^2 - p^2)} \right) \left[1 - Q_1\left(\frac{ab}{\sqrt{a^2 - p^2}}, c\sqrt{a^2 - p^2} \right) \right] \right.$$

$$\left. + \exp\left(\frac{p^2 c^2}{2} \right) Q_1(b, ac) - 1 \right\}, \, p \neq a$$

$$(B.25)$$

$$\int_0^c x \exp(p^2 x^2 / 2) Q_1(b, px) dx =$$

$$(B.26)$$

$$\frac{1}{p^2} \left[\frac{pc}{b} \exp\left(-\frac{b^2}{2} \right) I_1(bpc) + \exp\left(\frac{p^2 c^2}{2} \right) Q_1(b, pc) - 1 \right]$$

$$\int_0^\infty x^{2n-1} \exp(-p^2 x^2 / 2) Q_1(b, ax) dx = \frac{2^{n-1}(n-1)!}{p^{2n}}$$

$$\times \left[1 - \frac{a^2}{a^2 + p^2} \exp\left(-\frac{p^2 b^2}{2(a^2 + p^2)} \right) \sum_{k=0}^{n-1} \left(\frac{p^2}{a^2 + p^2} \right)^k L_k\left(-\frac{a^2 b^2}{2(a^2 + p^2)} \right) \right] \quad (B.27)$$

$$\int_0^\infty x^{2n-1} \exp(-p^2 x^2 / 2) Q_1(ax, b) dx =$$

$$\frac{2^{n-1}(n-1)!}{p^{2n}} \frac{a^2}{a^2+p^2} \exp\left(-\frac{p^2 b^2}{2(a^2+p^2)}\right) \quad \text{(B.28)}$$

$$\times \sum_{k=0}^{n-1} \varepsilon_k \left(\frac{p^2}{a^2+p^2}\right)^k L_k\left(-\frac{a^2 b^2}{2(a^2+p^2)}\right)$$

where

$$\varepsilon_K = \begin{cases} 1, & k < n-1 \\ 1+\dfrac{p^2}{a^2}, & k = n-1 \end{cases} \quad \text{(B.29)}$$

and $L_k(x)$ is the Laguerre polynomial which is the special case of the generalized Laguerre polynomial defined in (A.24) corresponding to $n = 0$, i.e.,

$$L_k(x) \triangleq L_k^{(0)}(x) = \sum_{i=0}^{k} \binom{k}{k-i} \frac{(-x)^i}{i!} \quad \text{(B.30)}$$

E. Q-Function with One Linear Argument, Bessel Functions, Exponentials and Powers of x

$$\int_0^\infty x \exp(-p^2 x^2 / 2) I_0(cx) Q_1(ax, b) dx = \frac{1}{p^2} \exp\left(\frac{c^2}{2p^2}\right) Q_1\left(\frac{ac}{p\sqrt{a^2+p^2}}, \frac{bp}{\sqrt{a^2+p^2}}\right) \quad \text{(B.31)}$$

$$\int_0^\infty x \exp(-p^2 x^2 / 2) I_0(cx) Q_1(b, ax) dx =$$

$$\frac{1}{p^2}\left[\exp\left(\frac{c^2}{2p^2}\right) Q_1\left(\frac{bp}{\sqrt{a^2+p^2}}, \frac{ac}{p\sqrt{a^2+p^2}}\right) \quad \text{(B.32)}\right.$$

$$\left. -\frac{a^2}{a^2+p^2} \exp\left(\frac{c^2-p^2 b^2}{2(a^2+p^2)}\right) I_0\left(\frac{abc}{a^2+p^2}\right)\right]$$

$$\int_0^\infty x I_0(cx) Q_1(b, ax) dx = \frac{1}{a^2} \exp\left(\frac{c^2}{2a^2}\right)\left[I_0\left(\frac{bc}{a}\right) + \frac{ab}{c}I_1\left(\frac{bc}{a}\right)\right] \quad \text{(B.33)}$$

$$\int_0^\infty x \exp(-p^2 x^2 / 2) I_1(bx) Q_1(c, ax) dx =$$

$$\frac{1}{b}\left[\exp\left(\frac{b^2}{2p^2}\right) Q_1\left(\frac{pc}{\sqrt{a^2 + p^2}}, \frac{ab}{p\sqrt{a^2 + p^2}}\right) - 1\right] \quad \text{(B.34)}$$

$$\int_0^\infty x \exp(-p^2 x^2 / 2) I_1(bx) Q_1(ax, c) dx =$$

$$\frac{1}{b}\left[\exp\left(\frac{b^2}{2p^2}\right) Q_1\left(\frac{ab}{p\sqrt{a^2 + p^2}}, \frac{pc}{\sqrt{a^2 + p^2}}\right)\right.$$

$$\left. -\exp\left(-\frac{c^2}{2}\right) Q_1\left(\frac{jb}{\sqrt{a^2 + p^2}}, \frac{jac}{\sqrt{a^2 + p^2}}\right)\right] \quad \text{(B.35)}$$

$$\int_0^\infty x \exp(-p^2 x^2 / 2) I_1(acx) Q_1(ax, c) dx =$$

$$\frac{1}{ac}\left\{\exp\left(\frac{a^2 c^2}{2p^2}\right) Q_1\left(\frac{a^2 c}{p\sqrt{a^2 + p^2}}, \frac{pc}{\sqrt{a^2 + p^2}}\right)\right.$$

$$\left. -\frac{1}{2}\exp\left(-\frac{c^2}{2}\right)\left[1 + \exp\left(\frac{a^2 c^2}{a^2 + p^2}\right) I_0\left(\frac{a^2 c^2}{a^2 + p^2}\right)\right]\right\} \quad \text{(B.36)}$$

Note that in checking the consistency between (B.35) and (B.36), use has been made of the relation

$$Q_1(\alpha, \alpha) = \frac{1}{2}\left[1 + \exp(-\alpha^2) I_0(\alpha^2)\right] \quad \text{(B.37)}$$

which holds even when α is imaginary and also the fact that $I_0(x) = I_0(-x)$.

$$\int_0^\infty I_1(cx) Q_1(b, ax) dx = \frac{1}{c}\left[\exp\left(\frac{c^2}{2a^2}\right) I_0\left(\frac{bc}{a}\right) - 1\right] \quad \text{(B.38)}$$

$$\int_0^\infty x^2 I_1(cx)Q_1(b,ax)dx = \frac{1}{a^4c^2}\exp\left(\frac{c^2}{2a^2}\right)\left[c(c^2+a^2b^2)I_0\left(\frac{bc}{a}\right)\right.$$

$$\left.+2ab(c^2-a^2)I_1\left(\frac{bc}{a}\right)\right] \tag{B.39}$$

$$\int_0^\infty x^2 \exp(-p^2x^2/2)I_1(cx)Q_1(ax,b)dx =$$

$$\frac{1}{p^4}\left[c\exp\left(\frac{c^2}{2p^2}\right)Q_1\left(\frac{ac}{p\sqrt{a^2+p^2}},\frac{bp}{\sqrt{a^2+p^2}}\right)\right. \tag{B.40}$$

$$\left.+\frac{abp^2}{a^2+p^2}\exp\left(\frac{c^2-b^2p^2}{2(a^2+p^2)}\right)I_1\left(\frac{abc}{a^2+p^2}\right)\right]$$

$$\int_0^\infty x^2 \exp(-p^2x^2/2)I_1(cx)Q_1(b,ax)dx =$$

$$\frac{1}{p^4}\left\{c\exp\left(\frac{c^2}{2p^2}\right)Q_1\left(\frac{bp}{\sqrt{a^2+p^2}},\frac{ac}{p\sqrt{a^2+p^2}}\right)\right.$$

$$-\frac{a^2}{(a^2+p^2)^2}\exp\left(\frac{c^2-b^2p^2}{2(a^2+p^2)}\right)\left[c(a^2+2p^2)\right. \tag{B.41}$$

$$\left.\left.\times I_0\left(\frac{abc}{a^2+p^2}\right)+abp^2I_1\left(\frac{abc}{a^2+p^2}\right)\right]\right\}$$

F. Product of Two Q-Functions with One Linear Argument

$$\int_0^\infty x\exp(-p^2x^2/2)Q_1(ax,b)Q_1(cx,d)dx = \frac{1}{p^2}\left\{\exp\left(-\frac{p^2}{a^2+p^2}\frac{b^2}{2}\right)\right.$$

$$\times\left[1-Q_1\left(\frac{d\sqrt{a^2+p^2}}{\sqrt{c^2+a^2+p^2}},\frac{abc}{\sqrt{a^2+p^2}\sqrt{c^2+a^2+p^2}}\right)\right] \tag{B.42}$$

$$\left.+\exp\left(-\frac{p^2}{c^2+p^2}\frac{d^2}{2}\right)Q_1\left(\frac{acd}{\sqrt{c^2+p^2}\sqrt{c^2+a^2+p^2}},\frac{b\sqrt{c^2+p^2}}{\sqrt{c^2+a^2+p^2}}\right)\right\}$$

G. Q-Function with Two Linear Arguments and x

$$\int_0^c xQ_1(ax,bx)dx = \frac{c^2}{2}Q_1(ac,bc) + \frac{bc^2}{2(a^2-b^2)}$$

$$\times \exp\left(-\frac{a^2+b^2}{2}c^2\right)\left[bI_0(abc^2) + aI_1(abc^2)\right]$$

(B.43)

$$-\frac{b^2}{|a^2-b^2|(a^2-b^2)}\left[1+\exp\left(-\frac{a^2+b^2}{2}c^2\right)I_0(abc^2)\right]$$

$$-2Q_1(c\min(a,b),c\max(a,b))], \quad a \neq b$$

$$\int_0^c xQ_1(ax,ax)dx = \frac{c^2}{4}\left\{1 + \exp(-a^2c^2)\left[I_0(a^2c^2) + I_1(a^2c^2)\right]\right\} \quad (B.44)$$

H. Q-Function with Two Linear Arguments, Exponentials and x

To simplify the notation, define

$$s = a^2 + b^2 + p^2, \quad t = a^2 - b^2 + p^2, \quad r = \sqrt{s^2 - 4a^2b^2} \qquad (B.45)$$

Then

$$\int_c^\infty x\exp(-p^2x^2/2)Q_1(ax,bx)dx = \frac{1}{p^2}\left[\exp\left(-\frac{p^2c^2}{2}\right)Q_1(ac,bc)\right.$$

$$\left. +\frac{t}{r}Q_1\left(c\sqrt{\frac{s-r}{2}},c\sqrt{\frac{s+r}{2}}\right) - \frac{1}{2}\left(1+\frac{t}{r}\right)\exp\left(-\frac{sc^2}{2}\right)I_0(abc^2)\right]$$

(B.46)

$$\int_0^c x\exp(-p^2x^2/2)Q_1(ax,bx)dx = \frac{1}{p^2}\left\{\frac{1}{2}\left(1+\frac{t}{r}\right)\left[1+\exp\left(-\frac{sc^2}{2}\right)I_0(abc^2)\right]\right.$$

$$\left. -\exp\left(-\frac{p^2c^2}{2}\right)Q_1(ac,bc) - \frac{t}{r}Q_1\left(c\sqrt{\frac{s-r}{2}},c\sqrt{\frac{s+r}{2}}\right)\right\}$$

(B.47)

$$\int_0^\infty x \exp(-p^2 x^2 /2) Q_1(ax, bx) dx = \frac{1}{2p^2}\left(1+\frac{t}{r}\right) \tag{B.48}$$

3. The Generalized (*mth*-Order) Marcum Q-Function

A. Q-Function with One Linear Argument and Powers of *x*

$$\int_0^\infty x^\mu Q_m(b, ax) dx = \frac{2^{(1+\mu)/2}\,\Gamma\!\left(m+\frac{1+\mu}{2}\right)}{(1+\mu)a^{1+\mu}\Gamma(m)}\,{}_1F_1\!\left(-\frac{1+\mu}{2}, m; -\frac{b^2}{2}\right),\ \mu>-1 \tag{B.49}$$

$$\int_0^\infty x^{2n-1} Q_m(b, ax) dx = \frac{2^{n-1}(n-1)!}{a^{2n}} L_n^{(m-1)}\!\left(-\frac{b^2}{2}\right),\ n\geq 1 \tag{B.50}$$

B. Q-Function with One Linear Argument, Exponentials and Powers of *x*

$$\int_0^\infty x^{2m-1} \exp(-p^2 x^2 /2) Q_m(ax, b) dx = \frac{2^{m-1}(m-1)!}{p^{2m}}\exp\!\left(-\frac{b^2 p^2}{2(a^2+p^2)}\right) \tag{B.51}$$

$$\times \sum_{k=0}^{n-1} \frac{1}{k!}\left(\frac{b^2 p^2}{2(a^2+p^2)}\right)^k$$

$$\int_0^\infty x^\mu \exp(-p^2 x^2 /2) Q_m(ax, b) dx = \int_0^\infty x^\mu \exp(-p^2 x^2 /2) Q_{m-1}(ax, b) dx$$

$$+ \frac{\Gamma\!\left(\frac{1+\mu}{2}\right)\left(\frac{b^2}{2}\right)^{m-1}}{2(m-1)!\left(\frac{a^2+p^2}{2}\right)^{(1+\mu)/2}}\exp\!\left(-\frac{b^2}{2}\right){}_1F_1\!\left(\frac{1+\mu}{2}, m; \frac{a^2 b^2}{2(a^2+p^2)}\right) \tag{B.52}$$

$$\int_0^\infty x \exp(-p^2 x^2 / 2) Q_m(ax, b) dx = \frac{1}{p^2} \exp\left(-\frac{b^2}{2}\right) \left\{ \left(\frac{a^2 + p^2}{a^2}\right)^{m-1} \right.$$

$$\times \left[\exp\left(\frac{a^2 b^2}{2(a^2 + p^2)}\right) - \sum_{k=0}^{m-2} \frac{1}{k!} \left(\frac{a^2 b^2}{2(a^2 + p^2)}\right)^k \right] + \sum_{k=0}^{m-2} \frac{1}{k!} \left(\frac{b^2}{2}\right)^k \right\}$$

(B.53)

$$\int_c^\infty x \exp(-p^2 x^2 / 2) Q_m(b, ax) dx = \frac{1}{p^2} \left[\exp\left(-\frac{p^2 c^2}{2}\right) Q_m(b, ac) - \left(\frac{a^2}{a^2 + p^2}\right)^m \right.$$

$$\times \exp\left(-\frac{b^2 p^2}{2(a^2 + p^2)}\right) Q_m\left(\frac{ab}{\sqrt{a^2 + p^2}}, c\sqrt{a^2 + p^2}\right) \right]$$

(B.54)

$$\int_c^\infty x \exp(p^2 x^2 / 2) Q_m(b, ax) dx = \frac{1}{p^2} \left[\left(\frac{a^2}{a^2 - p^2}\right)^m \exp\left(\frac{b^2 p^2}{2(a^2 - p^2)}\right) \right.$$

(B.55)

$$\times Q_m\left(\frac{ab}{\sqrt{a^2 - p^2}}, c\sqrt{a^2 - p^2}\right) - \exp\left(\frac{p^2 c^2}{2}\right) Q_m(b, ac) \right], a > p$$

$$\int_c^\infty x \exp(-p^2 x^2 / 2) Q_m(ax, b) dx = \frac{1}{p^2} \left\{ \exp\left(-\frac{p^2 c^2}{2}\right) Q_m(ac, b) \right.$$

$$+ \left(\frac{a^2 + p^2}{a^2}\right)^{m-1} \exp\left(-\frac{b^2 p^2}{2(a^2 + p^2)}\right) \left[1 - Q_m\left(c\sqrt{a^2 + p^2}, \frac{ab}{\sqrt{a^2 + p^2}}\right) \right] \right\}$$

(B.56)

C. Q-Function with One Linear Argument, Bessel Functions, Exponentials and Powers of x

$$\int_0^\infty x^m \exp(-p^2 x^2 / 2) I_{m-1}(cx) Q_m(ax, b) dx = \frac{1}{c} \left(\frac{c}{p^2}\right)^m \exp\left(\frac{c^2}{2p^2}\right)$$

(B.57)

$$\times Q_m\left(\frac{ac}{p\sqrt{a^2 + p^2}}, \frac{bp}{\sqrt{a^2 + p^2}}\right)$$

D. *Q-Function with Two Linear Arguments, Exponentials and Powers of x*

$$\int_c^\infty x\exp(-p^2x^2/2)Q_m(ax,bx)dx =$$

$$\frac{1}{p^2}\left[\exp\left(-\frac{p^2c^2}{2}\right)Q_1(ac,bc)+\frac{t}{r}Q_1\left(c\sqrt{\frac{s-r}{2}},c\sqrt{\frac{s+r}{2}}\right)\right.$$

$$-\frac{1}{2}\left(1+\frac{t}{r}\right)\exp\left(-\frac{sc^2}{2}\right)I_0(abc^2)\right]+\frac{1}{r}\sum_{k=1}^{m-1}\frac{1}{a^{2k}}\left\{\left(\frac{s-r}{2}\right)^k\right.$$

$$\times Q_k\left(c\sqrt{\frac{s-r}{2}},c\sqrt{\frac{s+r}{2}}\right)+\left(\frac{s+r}{2}\right)^k\left[1-Q_k\left(c\sqrt{\frac{s+r}{2}},c\sqrt{\frac{s-r}{2}}\right)\right]\right\}$$

$$(\text{B.58})$$

where r, s, and t are defined in (B.45).

$$\int_0^\infty x\exp(-p^2x^2/2)Q_m(ax,bx)dx = \frac{1}{2p^2}\left(1+\frac{t}{r}\right)+\frac{1}{r}\left[\frac{\frac{s-r}{2a^2}-\left(\frac{s-r}{2a^2}\right)^m}{1-\frac{s-r}{2a^2}}\right] \quad (\text{B.59})$$

$$\int_0^\infty x^{2k-1}\exp(-p^2x^2/2)Q_m(ax,bx)dx = \frac{1}{2}\left(\frac{2}{p^2}\right)^k(k-1)!\left\{1+\frac{b^{2m}}{s^m}\sum_{l=0}^{k-1}\binom{m+l}{l}\right.$$

$$\times\left(\frac{p^2}{s}\right)^l\left[\frac{a^2}{s}\,_2F_1\left(\frac{m+l+1}{2},\frac{m+l+2}{2};m+1;\frac{4a^2b^2}{s^2}\right)\right.$$

$$\left.-\left(\frac{m}{m+l}\right)_2F_1\left(\frac{m+l}{2},\frac{m+l+1}{2};m;\frac{4a^2b^2}{s^2}\right)\right]\right\}$$

$$(\text{B.60})$$

For the special case $k = m = 1$, using the relations

$$_2F_1\left(1,\frac{3}{2};2;z\right) = \frac{2}{\sqrt{1-z}}\left(\frac{1}{1+\sqrt{1-z}}\right) = \frac{2}{z}\left(\frac{1}{\sqrt{1-z}}-1\right) \quad (\text{B.61})$$

and

$$_2F_1\left(\frac{1}{2},1;1;z\right) = \frac{1}{\sqrt{1-z}} \tag{B.62}$$

the result in (B.60) reduces to the simple form in (B.48).

APPENDIX C: BOUNDS ON THE GAUSSIAN Q-FUNCTION AND THE MARCUM Q-FUNCTION

To facilitate evaluations involving the Gaussian Q-function and the Marcum Q-function, it is quite often useful to have upper and lower bounds on these functions as an alternative to their exact integral representations. In this appendix, we present a variety of such bounds many of which are derived from the alternative forms of the functions presented in Appendix B. Our interest here is not in comparing the degree of tightness of these various bounds but rather to merely present them to the reader allowing him or her to determine the advantages and disadvantages of each form for his or her particular application.

1. The Gaussian Q-Function

Perhaps the simplest upper bound on the Gaussian Q-function can be immediately obtained from the alternative form given in (A.1) by noting that the integrand is a monotonically increasing function of the integration variable and thus the maximum of this integrand occurs at $\theta = \pi / 2$. Thus, upper bounding the integrand by its maximum value, namely, $\exp(-x^2 / 2)$, we immediately obtain

$$Q(x) \le \frac{1}{2} \exp\left(-\frac{x^2}{2}\right) \tag{C.1}$$

which is equivalent to the Chernoff bound on this function improved by a factor of $1/2$. This procedure for obtaining the upper bound can be extended to provide tighter bounds [31]. In particular, we note that any monotonically increasing function $f(x)$ in the interval $\left[x_{\min}, x_{\max}\right]$

can be upper bounded by a piecewise constant (staircase) function $\hat{f}(x)$ defined as follows:

$$\hat{f}(x) = \begin{cases} f(x_1), x_{min} = x_0 \leq x \leq x_1 \\ f(x_2), x_1 < x \leq x_2 \\ \vdots \\ f(x_{N-1}), x_{N-2} < x \leq x_{N-1} \\ f(x_{max}), x_{N-1} < x \leq x_N = x_{max} \end{cases} \qquad (C.2)$$

where $x_1 < x_2 < ... < x_{N-1}$ is a set of $N-1$ monotonically increasing values of x arbitrarily chosen in the interval (x_{min}, x_{max}). Since the integrand of (A.1) is a monotonically increasing function of θ, then it can be upper bounded in the interval $(-\pi/2, \pi/2)$ as in (C.2) in which case the integral, $Q(x)$, is upper bounded by

$$Q(x) \leq \frac{1}{\pi} \sum_{i=1}^{N} (\theta_i - \theta_{i-1}) \exp\left(-\frac{x^2}{2\sin^2 \theta_i}\right) \qquad (C.3)$$

where $0 = \theta_0 < \theta_1 < \theta_2 < ... < \theta_{N-1} < \theta_N = \pi/2$. Note that (C.3) is nothing other than a particular form of Riemann sum that is ordinarily used to approximate the integral of a function, and thus as N increases, the tightness of the bound improves. Clearly, for the special case $N = 1$ whereby $\theta_0 = 0$ and $\theta_1 = \pi/2$, (C.3) reduces to (C.1).

The two-dimensional Gaussian Q-function of (A.39) can similarly be upper bounded. Consider first the case where $1 - \rho y/x$ (or $1 - \rho x/y$) is positive. Then, as for the one-dimensional Gaussian Q-function, the integrands of (A.39) are a monotonically increasing function of θ in their respective integration intervals in which case the integral, $Q(x, y:\rho)$, is upper bounded by

$$Q(x, y:\rho) \leq \frac{1}{2\pi} \sum_{i=1}^{N_{I_1}-1} (\theta_i - \theta_{i-1}) \exp\left(-\frac{x^2}{2\sin^2 \theta_i}\right) + \frac{1}{2\pi} (\theta_1^* - \theta_{N_{I_1}}) \exp\left(-\frac{x^2}{2\sin^2 \theta_1^*}\right)$$

$$+ \frac{1}{2\pi} \sum_{i=1}^{N_{I_2}-1} (\theta_i - \theta_{i-1}) \exp\left(-\frac{y^2}{2\sin^2 \theta_i}\right) + \frac{1}{2\pi} (\theta_2^* - \theta_{N_{I_2}}) \exp\left(-\frac{y^2}{2\sin^2 \theta_2^*}\right),$$

$$x \geq 0, y \geq 0$$

$$(C.4)$$

where

$$\theta_1^* = \tan^{-1}\left(\frac{\sqrt{1-\rho^2}\,y/x}{1-\rho y/x}\right), \theta_2^* = \tan^{-1}\left(\frac{\sqrt{1-\rho^2}\,x/y}{1-\rho x/y}\right) \tag{C.5}$$

and N_{l_1}, N_{l_2} are the smallest integers such that

$$\theta_{N_{l_1}} \geq \theta_1^*, \theta_{N_{l_2}} \geq \theta_2^* \tag{C.6}$$

For the case where $1-\rho y/x$ (or $1-\rho x/y$) is negative, then in the interval between $\pi/2$ and the upper limit, the integrands are a monotonically decreasing function of θ and can be upper bounded by a downward staircase function analogous to (C.2). Hence, for this case, defining N_{u_1}, N_{u_2} as the largest integers such that

$$\theta_{N_{u_1}} \leq \theta_1^*, \theta_{N_{u_2}} \leq \theta_2^* \tag{C.7}$$

then the two-dimensional Gaussian Q-function can be upper bounded by

$$Q(x,y:\rho) \leq \frac{1}{2\pi}\sum_{i=1}^{N}(\theta_i - \theta_{i-1})\exp\left(-\frac{x^2}{2\sin^2\theta_i}\right) + \frac{1}{2\pi}\sum_{i=N_{u_1}+2}^{N}(\theta_i - \theta_{i-1})$$

$$\times \exp\left(-\frac{x^2}{2\sin^2\theta_i}\right) + \frac{1}{2\pi}\left[\theta_{N_{u_1}+1} - (\pi - \theta_1^*)\right]\exp\left(-\frac{x^2}{2\sin^2\theta_{N_{u_1}+1}}\right)$$

$$+ \frac{1}{2\pi}\sum_{i=1}^{N}(\theta_i - \theta_{i-1})\exp\left(-\frac{y^2}{2\sin^2\theta_i}\right) + \frac{1}{2\pi}\sum_{i=N_{u_2}+2}^{N}(\theta_i - \theta_{i-1}) \tag{C.8}$$

$$\times \exp\left(-\frac{y^2}{2\sin^2\theta_i}\right) + \frac{1}{2\pi}\left[\theta_{N_{u_2}+1} - (\pi - \theta_2^*)\right]\exp\left(-\frac{y^2}{2\sin^2\theta_{N_{u_2}+1}}\right),$$

$$x \geq 0, y \geq 0$$

Another interesting upper bound related to the ability of separating the argument of the function can be obtained from the alternative form in (A.1). In particular, we first note from this equation that

$$Q(\sqrt{x_1 + x_2}) = \frac{1}{\pi}\int_0^{\pi/2} \exp\left(-\frac{x_1}{2\sin^2\theta}\right)\exp\left(-\frac{x_2}{2\sin^2\theta}\right)d\theta \qquad (C.9)$$

Now applying Schwarz' inequality [32] to the integral in (C.9), we have

$$Q(\sqrt{x_1 + x_2}) \leq \sqrt{\left[\frac{1}{\pi}\int_0^{\pi/2}\exp\left(-\frac{x_1}{\sin^2\theta}\right)d\theta\right]\left[\frac{1}{\pi}\int_0^{\pi/2}\exp\left(-\frac{x_2}{\sin^2\theta}\right)d\theta\right]} \qquad (C.10)$$

$$= \sqrt{Q(\sqrt{2x_1})}\sqrt{Q(\sqrt{2x_2})}$$

A simple integral form of the deviation of the Gaussian Q-function from the first term of its asymptotic series can be obtained from the alternative form of this function in (A.36). First integrating (A.36) gives

$$\int_0^x Q(y)dy = \frac{1}{\sqrt{2\pi}}\left[1 - \int_0^{\pi/2}\cos\theta\exp\left(-\frac{x^2}{2\sin^2\theta}\right)d\theta\right] \qquad (C.11)$$

But the definite integral of the Gaussian Q-function can be found in closed form using integration by parts. In particular,

$$\int_0^x Q(y)dy = xQ(x) - \frac{1}{\sqrt{2\pi}}\left[\exp\left(-\frac{x^2}{2}\right) - 1\right] \qquad (C.12)$$

Equating (C.11) and (C.12) gives

$$Q(x) = \frac{1}{\sqrt{2\pi}x}\exp\left(-\frac{x^2}{2}\right) - \frac{1}{\sqrt{2\pi}x}\int_0^{\pi/2}\cos\phi\exp\left(-\frac{x^2}{2\sin^2\phi}\right)d\phi \qquad (C.13)$$

where the first term corresponds to the leading term in the asymptotic expansion of $Q(x)$. Thus, the second term can be regarded as an integral representation of the deviation of $Q(x)$ from its approximation by the first term of its asymptotic series. Furthermore, since the second term is always negative for $x > 0$, then the first term is also an upper bound on $Q(x)$, i.e.,

$$Q(x) < \frac{1}{\sqrt{2\pi}x}\exp\left(-\frac{x^2}{2}\right) \qquad (C.14)$$

A lower bound on $Q(x)$ can be obtained by first rewriting the classical representation of $Q(x)$ in (1.3) as

$$Q(x) \triangleq \frac{1}{\sqrt{2\pi}} \int_x^\infty \frac{1}{y} y \exp\left(-\frac{y^2}{2}\right) dy \qquad (C.15)$$

and then integrating by parts using $\int u\, dv = uv - \int v\, du$ where

$$u = \frac{1}{y}, \quad du = -\frac{1}{y^2} dy$$

$$v = -\exp\left(-\frac{y^2}{2}\right), \quad dv = y \exp\left(-\frac{y^2}{2}\right) dy \qquad (C.16)$$

resulting in

$$Q(x) = \frac{1}{\sqrt{2\pi}} \left[\frac{1}{x} \exp\left(-\frac{x^2}{2}\right) - \int_x^\infty \frac{1}{y^2} \exp\left(-\frac{y^2}{2}\right) dy \right] \qquad (C.17)$$

Further integrating by parts the second term in (C.17) with

$$u = \frac{1}{y^3}, \quad du = -\frac{3}{y^4} dy$$

$$v = -\exp\left(-\frac{y^2}{2}\right), \quad dv = y \exp\left(-\frac{y^2}{2}\right) dy \qquad (C.18)$$

we obtain

$$Q(x) = \frac{1}{\sqrt{2\pi}} \left[\frac{1}{x} \exp\left(-\frac{x^2}{2}\right) - \frac{1}{x^3} \exp\left(-\frac{x^2}{2}\right) + \int_x^\infty \frac{3}{y^4} \exp\left(-\frac{y^2}{2}\right) dy \right] \quad (C.19)$$

Ignoring the third (integral) term in (C.19), we obtain the desired lower bound for $x > 0$

$$Q(x) > \frac{1}{\sqrt{2\pi}x} \left(1 - \frac{1}{x^2}\right) \exp\left(-\frac{x^2}{2}\right) \qquad (C.20)$$

2. The Marcum Q-Function

Simple upper and lower bounds on $Q_1(\alpha, \beta)$ can be obtained in the same manner that the upper bound on the Gaussian Q-function in (C.1) was obtained from (A.1). In particular, for $\beta > \alpha \geq 0$, we observe that the maximum and minimum of the integrand in (A.6) occurs for $\theta = -\pi/2$ and $\theta = \pi/2$, respectively. Thus, replacing the integrand by its maximum and minimum values leads to the upper and lower "Chernoff-type" bounds

$$\frac{\beta}{\beta + \alpha} \exp\left[-\frac{(\beta + \alpha)^2}{2}\right] \leq Q_1(\alpha, \beta) \leq \frac{\beta}{\beta - \alpha} \exp\left[-\frac{(\beta - \alpha)^2}{2}\right] \quad (C.21)$$

which, in view of the fact that $Q_1(0, \beta) = \exp(-\beta^2/2)$ are asymptotically tight as $\alpha \to 0$.

For $\alpha > \beta \geq 0$, the integrand in (A.7) has a minimum at $\theta = -\pi/2$ and a maximum at $\theta = \pi/2$. Since the maximum of the integrand, $[\beta/(\alpha + \beta)]\exp[-(\alpha + \beta)^2/2]$, is always positive, the upper bound obtained by replacing the integrand by this value would exceed unity and hence be useless. On the other hand, the minimum of the integrand, $-[\beta/(\alpha - \beta)]\exp[-(\alpha - \beta)^2/2]$, is always negative. Hence a lower Chernoff-type bound on $Q_1(\alpha, \beta)$ is given by[5]

$$1 - \frac{\alpha}{\alpha - \beta} \exp\left[-\frac{(\alpha - \beta)^2}{2}\right] \leq Q_1(\alpha, \beta) \quad (C.22)$$

The forms of the Marcum Q-function in (A.10) and (A.11) immediately allow obtaining tighter upper and lower bounds of this function than those in (C.21) and (C.22). In particular, once again recognizing that for $\beta > \alpha \geq 0$ the maximum and minimum of the first exponential integrand in (A.10) occurs for $\theta = -\pi/2$ and $\theta = \pi/2$, respectively, and vice versa for the second exponential integrand, then we immediately obtain

$$\exp\left[-\frac{(\beta + \alpha)^2}{2}\right] \leq Q_1(\alpha, \beta) \leq \exp\left[-\frac{(\beta - \alpha)^2}{2}\right] \quad (C.23)$$

---

[5] Clearly since $Q_1(\alpha, \beta)$ can never be negative, the lower bound is only useful for values of the arguments that result in a non-negative value.

Making a similar recognition in (A.11), then for $\alpha > \beta \geq 0$ we obtain the lower bound

$$1 - \frac{1}{2}\left\{\exp\left[-\frac{(\alpha - \beta)^2}{2}\right] - \exp\left[-\frac{(\alpha + \beta)^2}{2}\right]\right\} \leq Q_1(\alpha, \beta) \qquad \text{(C.24)}$$

The advantage of the bounds is that they are simple and of exponential-type; however, they may not necessarily be the tightest bounds achievable over all ranges of their arguments. In this regard, the following upper and lower bounds have also been reported [33].

$$\exp\left(-\frac{\alpha^2 + \beta^2}{2}\right)I_0(\alpha\beta) \leq Q_1(\alpha, \beta) \leq \exp\left(-\frac{\alpha^2 + \beta^2}{2}\right)I_0(\alpha\beta) + \alpha\sqrt{\frac{\pi}{2}}Q(\beta - \alpha)$$

$$\text{(C.25)}$$

For $\beta > \alpha$, the bounds in (C.25) are tighter than those in (C.23). However, for $\alpha > \beta$ the lower bound in (C.24) is very tight and better than that in (C.25).

Upper and lower bounds on the mth-order Marcum Q-function are not readily obtainable from the alternative forms in (A.16) and (A.17). Nevertheless, it is possible [34] to obtain such bounds by using the upper and lower bounds on the first-order Marcum Q-function given in (C.23) together with the recursive relation

$$Q_m(\alpha, \beta) = \exp\left(-\frac{\alpha^2 + \beta^2}{2}\right)\sum_{n=1}^{m-1}\left(\frac{\beta}{\alpha}\right)^n I_n(\alpha\beta) + Q_1(\alpha, \beta) \qquad \text{(C.26)}$$

The results are as follows:

$$Q_m(\alpha, \beta) \leq \exp\left[-\frac{(\beta - \alpha)^2}{2}\right] + \frac{1}{\pi}\left\{\exp\left[-\frac{(\beta - \alpha)^2}{2}\right] - \exp\left[-\frac{(\beta + \alpha)^2}{2}\right]\right\}$$

$$\times \left(\frac{\beta}{\alpha}\right)^{m-1}\left[\frac{1 - (\alpha/\beta)^{m-1}}{1 - \alpha/\beta}\right] \qquad \text{(C.27)}$$

$$\exp\left(-\frac{(\beta + \alpha)^2}{2}\right)\sum_{n=0}^{m-1}\frac{(\beta/2)^n}{n!} \leq Q_m(\alpha, \beta), \quad 0 \leq \alpha < \beta \qquad \text{(C.28)}$$

$$1 - \frac{1}{2}\left\{ \exp\left[-\frac{(\alpha - \beta)^2}{2} \right] - \exp\left[-\frac{(\alpha + \beta)^2}{2} \right] \right\}$$

$$+ \exp\left(-\frac{(\alpha + \beta)^2}{2} \right) \sum_{n=1}^{m-1} \frac{(\beta / 2)^n}{n!} \leq Q_m(\alpha, \beta), 0 \leq \beta < \alpha$$

(C.29)

Note that the first term in (C.27) represents the upper bound on the first-order Marcum Q-function as given in (C.23). Similarly, the first term (corresponding to $n = 0$) in (C.28) is the lower bound on the first-order Marcum Q-function as given in (C.23). Also, for $\alpha = 0$, (C.28) becomes equal to the exact result for $Q_m(0, \beta)$ as given by

$$Q_m(0, \beta) = \sum_{n=0}^{m-1} \exp\left(-\frac{\beta^2}{2} \right) \frac{(\beta^2 / 2)^n}{n!}$$

(C.30)

Finally, because of the monotonicity of the integrands in (A.12) and (A.13), the technique used to upper bound the first- and second-order Gaussian Q-functions in (C.3) and (C.4) can also used to upper bound the first-order Marcum Q-function. In particular, the following results are obtained:

$$Q_1(\alpha, \beta) \leq \frac{1}{2\pi} \sum_{i=1}^{N} (\theta_i - \theta_{i-1}) \left\{ \exp\left[-\frac{\beta^2}{2}\left(1 + 2\left(\frac{\alpha}{\beta}\right)\cos\theta_i + \left(\frac{\alpha}{\beta}\right)^2 \right) \right] \right.$$

$$\left. + \exp\left[-\frac{\beta^2}{2}\left(1 - \left(\frac{\alpha}{\beta}\right)^2 \right)^2 \left(1 + 2\left(\frac{\alpha}{\beta}\right)\cos\theta_{i-1} + \left(\frac{\alpha}{\beta}\right)^2 \right)^{-1} \right] \right\} d\theta,$$

$$\beta \geq \alpha \geq 0$$

(C.31)

and

$$Q_1(\alpha, \beta) \leq 1 + \frac{1}{2\pi} \sum_{i=1}^{N} (\theta_i - \theta_{i-1}) \left\{ \exp\left[-\frac{\alpha^2}{2}\left(1 + 2\left(\frac{\beta}{\alpha}\right)\cos\theta_i + \left(\frac{\beta}{\alpha}\right)^2 \right) \right] \right.$$

$$\left. - \exp\left[-\frac{\alpha^2}{2}\left(1 - \left(\frac{\beta}{\alpha}\right)^2 \right)^2 \left(1 + 2\left(\frac{\beta}{\alpha}\right)\cos\theta_i + \left(\frac{\beta}{\alpha}\right)^2 \right)^{-1} \right] \right\} d\theta,$$

$$\alpha \geq \beta \geq 0$$

(C.32)

REFERENCES

[1] I. S. Gradshteyn and I. M. Rhyzhik, *Table of Integrals, Series, and Products*, 5th ed., San Diego, CA: Academic Press, 1994.

[2] M. Abramowitz and I. A. Stegun, *Handbook of Mathematic Functions with Formulas, Graphs, and Mathematical Tables*, U. S. Department of Commerce, National Bureau of Standards Applied Mathematics Series #55 (reprinted by Dover Press), 9th printing, 1970.

[3] G. E. Andrews, R. Askey and R. Roy, *Special Functions*, Cambridge University Press.

[4] L. B. W. Jolley, *Summation of Series*, 2nd revised ed., New York, NY: Dover Press.

[5] K. S. Miller, *Multidimensional Gaussian Distributions*, New York, NY: John Wiley & Sons, Inc., 1964.

[6] J. K. Omura and T. Kailath "Some Useful Probability Functions" Tech. Rep. 7050-6, Stanford University, September 1965.

[7] M. Nakagami, "The m-distribution: a general formula of intensity distribution of rapid fading," in *Statistical Methods in Radio Wave Propagation*. Oxford: Pergamon Press, 1960, pp. 3–36.

[8] J. I. Marcum, Table of Q Functions, U.S. Air Force Project RAND Research Memorandum M-339, ASTIA Document AD 1165451, Rand Corporation, Santa Monica, CA, January 1, 1950.

[9] A. H. Nuttall, "Some integrals involving the Q-function," Technical Report TR4297, Naval Underwater Systems Center, Newport, RI, April 17, 1972.

[10] National Bureau of Standards, *Tables of the Bivariate Normal Distribution Function and Related Functions*, Appl. Math. Series no. 50, Washington, D.C.: U. S. Government Printing Office, Superintendant of Documents, 1959.

[11] R. Price, "Some non-central F-distributions expressed in closed form," *Biometrika*, vol. 51, issue 1/2, June 1964, pp. 107–122.

[12] H. A. David, *Order Statistics*, New York: John Wiley & Sons, Inc., 1981.

[13] B. C. Arnold, N. Balakrishnan, and H. N. Nagaraja, *A First Course in Order Statistics*, New York: John Wiley & Sons, Inc., 1992.

[14] J. Proakis, *Digital Communications*, New York, NY: McGraw-Hill, 4th ed., 2001.

[15] M. K. Simon and M.-S. Alouini, *Digital Communications over Fading Channels: A Unified Approach to Performance Analysis*, New York, NY: John Wiley & Sons, Inc., 2000.

[16] M.K. Simon and M.-S. Alouini, "On the difference of two chi-squared variates with application to outage probability computation," *IEEE Trans. Commun.*, vol. 49, no. 11, November 2001, pp. 1946–1954. Also see *38th Annual Allerton Conf. Rec.*, Monticello, IL, October 2000, pp. 42–51.

[17] G. Turin, "The characteristic function of Hermetian quadratic forms in complex normal variables," *Biometrika*, vol. 47, issue 1/2, June 1960, pp. 199–201.

[18] F. S. Weinstein, "Simplified relationships for the probability distribution of the phase of a sine wave in narrow-band normal noise," *IEEE Trans. Inf. Theory*, vol. IT-20, September 1974, pp. 658–661.

[19] R. F. Pawula, S. O. Rice and J. H. Roberts, "Distribution of the phase angle between two vectors perturbed by Gaussian noise," *IEEE Trans. Comm.*, vol. 30, August 1982, pp. 1828–1841.

[20] J. W. Craig, "A new, simple and exact result for calculating the probability of error for two-dimensional signal constellations," *IEEE MILCOM'91 Conf. Rec.*, Boston, pp. 25.5.1–25.5.5.

[21] K. Lever, "New derivation of Craig's formula for the Gaussian probability function," *Electron. Lett.*, vol. 34, September 1998, pp. 1821–1822.

[22] J. I. Marcum and P. Swerling, "Studies of target detection by pulsed radar," *IEEE Trans. Inf. Theory*, vol. IT-6, April 1960.

[23] S. Stein, "The Q function and related integrals," Research Report No. 467, Applied Research Laboratory, Sylvania Electronic Systems, Waltham, MA, June 29, 1965.

[24] C. W. Helstrom, *Elements of Signal Detection and Estimation*, Upper Saddle River, NJ: Prentice Hall, 1995.

[25] M. K. Simon, "A new twist on the Marcum Q-function and its application," *IEEE Commun. Lett.*, vol. 2, February 1998, pp. 39–41.

[26] R. F. Pawula, "Relations between the Rice Ie-function and Marcum Q-function with applications to error rate calculations," *Electron. Lett.*, vol. 31, September 28, 1995, pp. 1717–1719.

[27] R. F. Pawula, "A new formula for MDPSK symbol error probability," *IEEE Commun. Lett.*, vol. 2, October 1998, pp. 271–272.

[28] M. K. Simon, "A simpler form of the Craig representation for the two-dimensional Gaussian Q-function," *IEEE Comm. Lett.*, vol. 6, no. 2, February 2002, pp. 49–51.

[29] A. H. Nuttall, "Some integrals involving the Q_M-function," Technical Report TR4755, Naval Underwater Systems Center, Newport, RI, May 15, 1974.

[30] M.K. Simon and M.-S. Alouini, "Some new results for integrals involving the generalized Marcum Q-function and their application to performance evaluation over fading channels," to appear in *IEEE Transactions on Wireless Communications*, 2002.

[31] M. Chiani, D. Dardari, "Improved exponential bounds and approximation for the Q-function and related bounds on its inverse," to be published.

[32] A. Papoulis, *Signal Analysis*, New York: McGraw-Hill, Inc., 1977.

[33] M. Chiani, "Integral representation and bounds for Marcum Q-function," *Electron. Lett.*, March 1999, pp. 445–446.

[34] M. K. Simon and M.-S. Alouini, "Exponential-type bounds on the generalized Marcum Q-function with application to fading channel error probability analysis," *IEEE Trans. Commun.*, vol. 48, no. 3, March 2000, pp. 359–366.

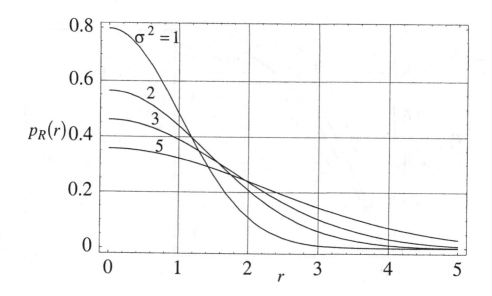

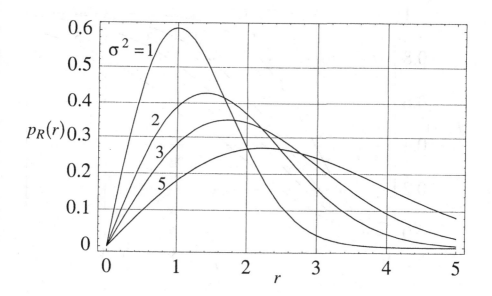

Fig. 1. Rayleigh PDFs: (a) $n=1$, Eq. (2.2); (b) $n=2$, Eq. (2.5).

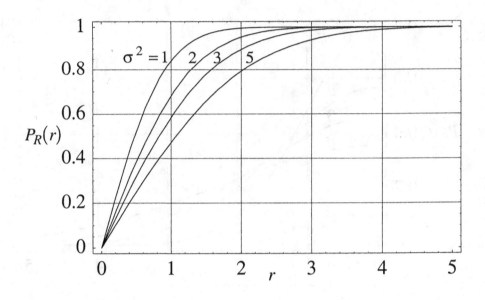

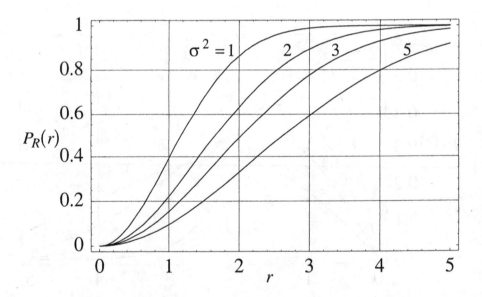

Fig. 2. Rayleigh: (a) $n=1$, Eq. (2.3); (b) $n=2$, Eq. (2.6).

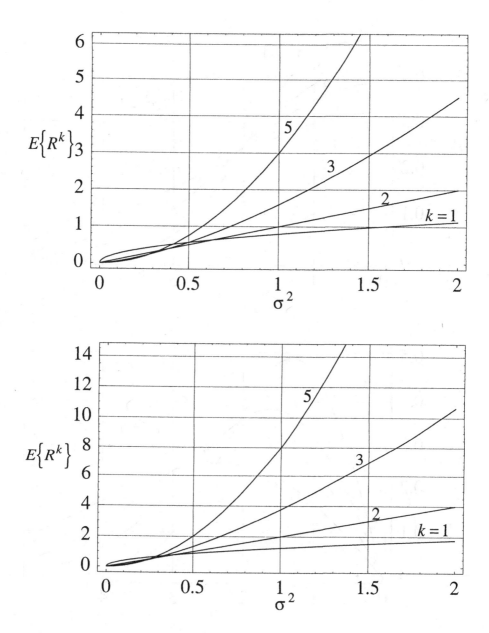

Fig. 3. Rayleigh Moments: (a) $n=1$, Eq. (2.4); (b) $n=2$, Eq. (2.7).

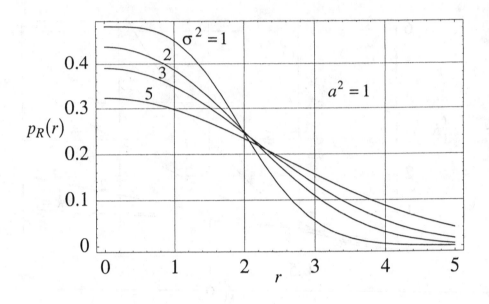

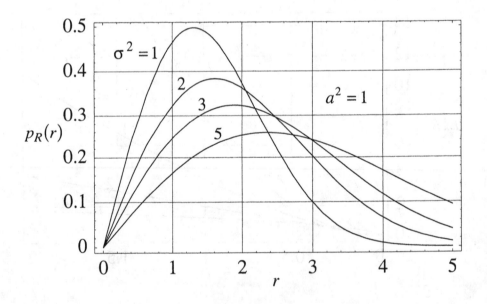

Fig. 4. Rician PDFs: (a) n=1, Eq. (2.14); (b) n=2, Eq. (2.17).

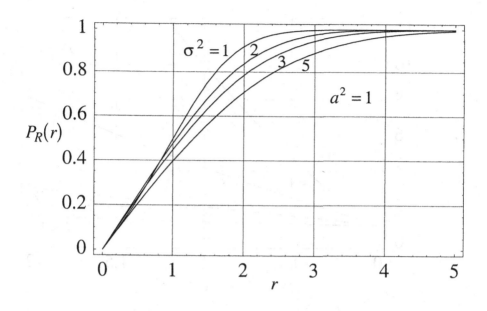

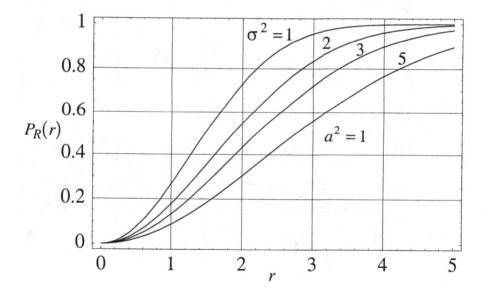

Fig. 5. Rician CDFs: (a) $n=1$, Eq. (2.15); (b) $n=2$, Eq. (2.18).

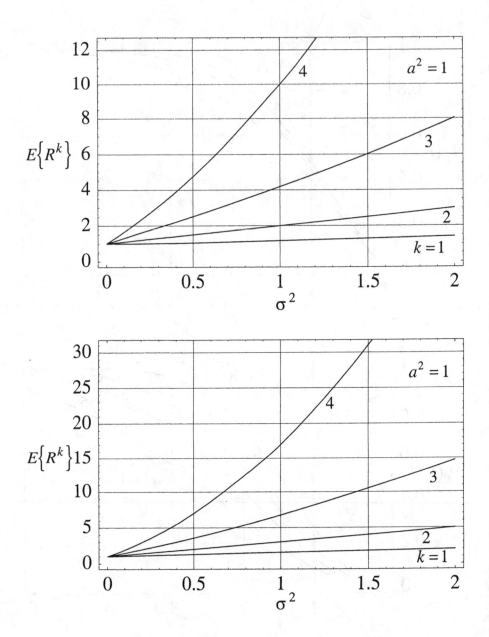

Fig. 6. Rician Moments: (a) $n=1$, Eq. (2.16); (b) $n=2$, Eq. (2.19).

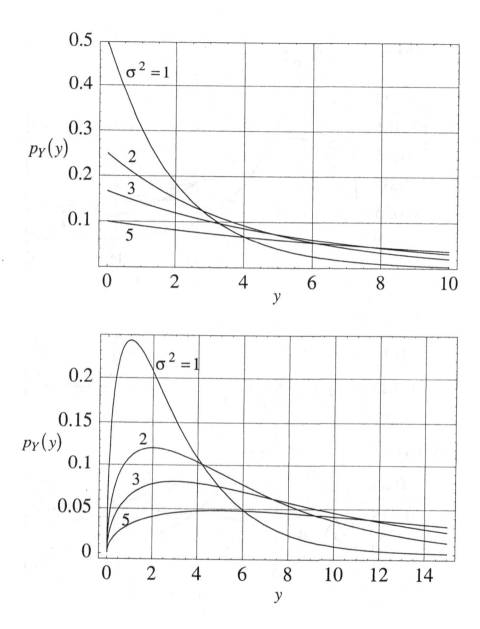

Fig. 7. Central Chi-Square PDFs: (a) $n=2$, Eq. (2.32); (b) $n=3$, Eq. (2.36); (c) $n=4$, Eq. (2.32); (d) $n=5$, Eq. (2.36).

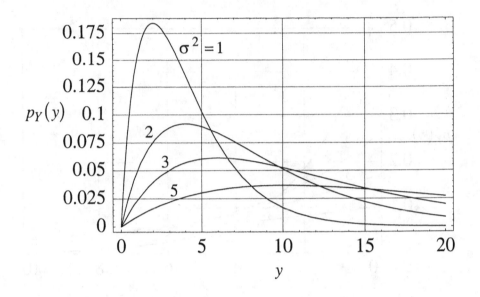

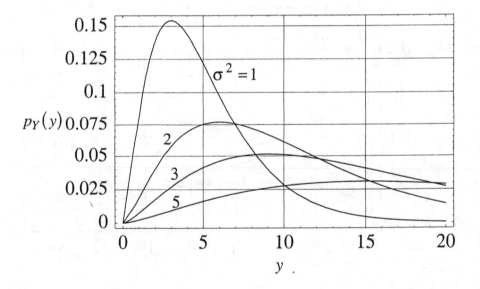

Fig. 7. cont'd.

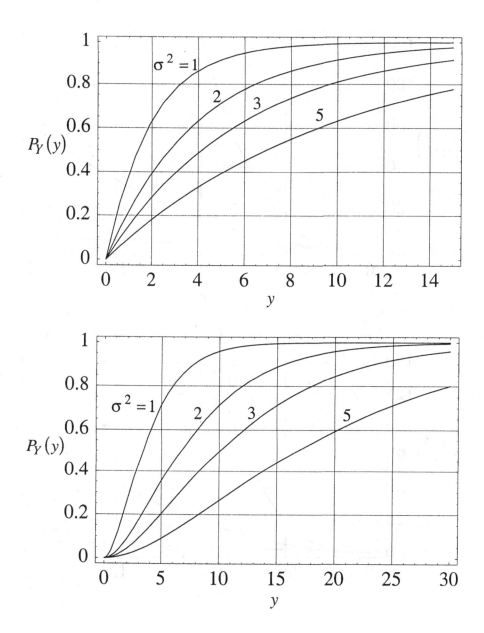

Fig. 8. Central Chi-Square CDFs: (a) $n=2$, Eq. (2.33); (b) $n=4$, Eq. (2.33).

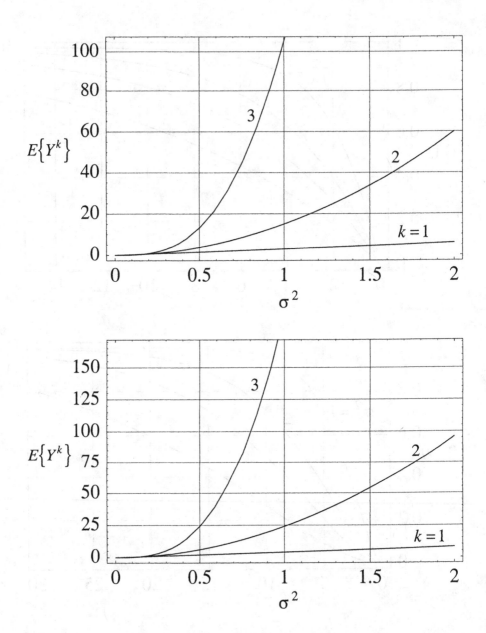

Fig. 9. Central Chi-Square Moments: (a) $n=3$, Eq. (2.39); (b) $n=4$, Eq. (2.34).

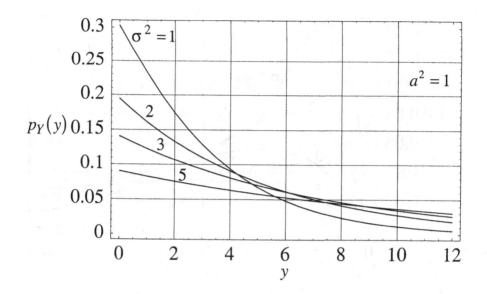

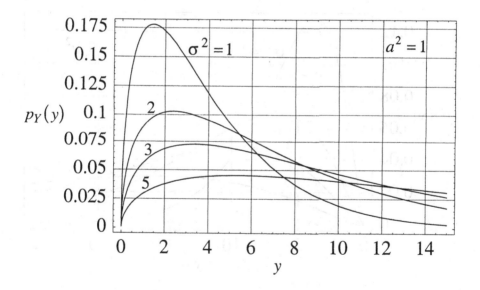

Fig. 10. Noncentral Chi-Square PDFs: (a) $n=2$, Eq. (2.44); (b) $n=3$, Eq. (2.48); (c) $n=4$, Eq. (2.44); (d) $n=5$, Eq. (2.48).

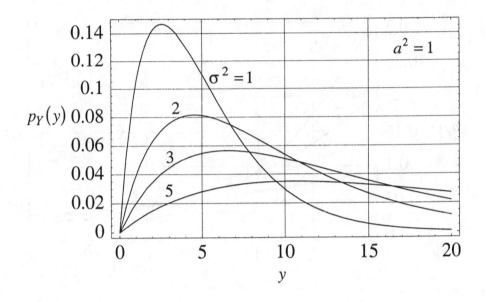

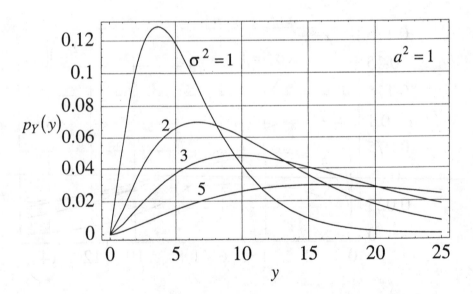

Fig. 10. cont'd.

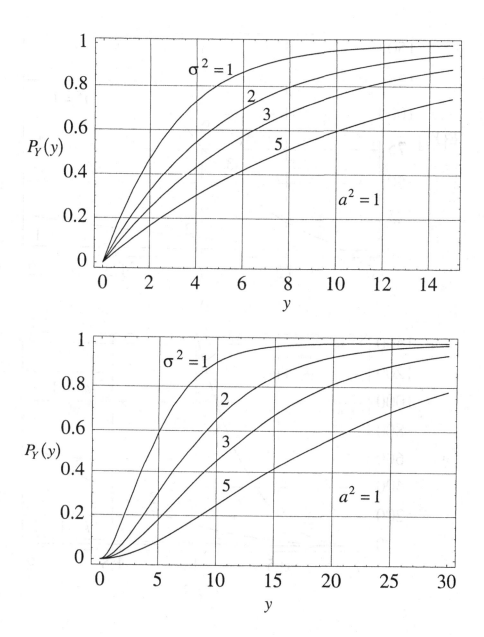

Fig. 11. Noncentral Chi-Square CDFs: (a) $n=2$, Eq. (2.45); (b) $n=4$, Eq. (2.45).

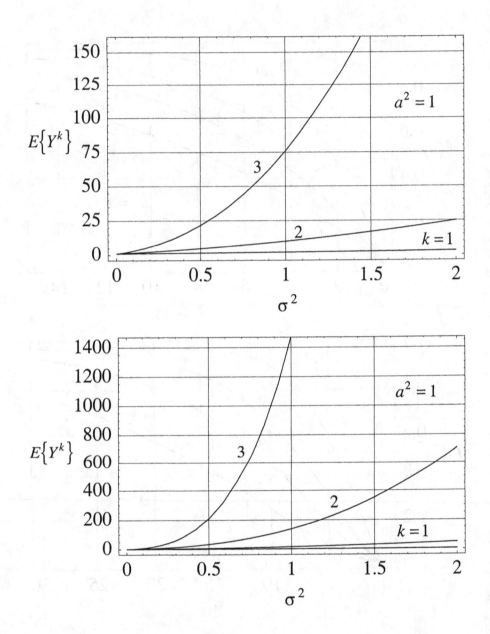

Fig. 12. Noncentral Chi-Square Moments: (a) n=1, Eq. (2.43); (b) n=2, Eq. (2.47).

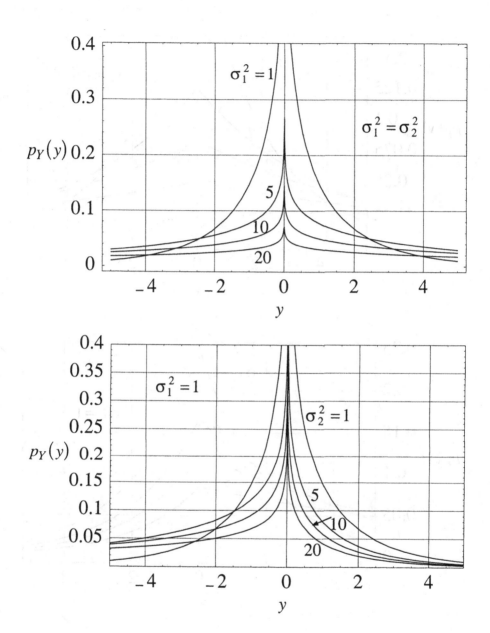

Fig. 13. Independent Central Chi-Square (-) Central Chi-Square PDFs: (a) $n_1=n_2=1$, equal variance, Eq. (4.1); (b) $n_1=n_2=1$, unequal variance, Eq. (4.1); (c) $n_1=n_2=2$, equal variance, Eq. (4.4); (d) $n_1=n_2=2$, unequal variance, Eq. (4.4).

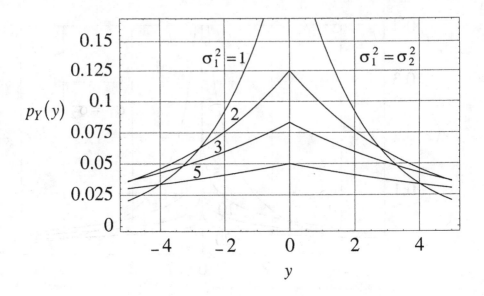

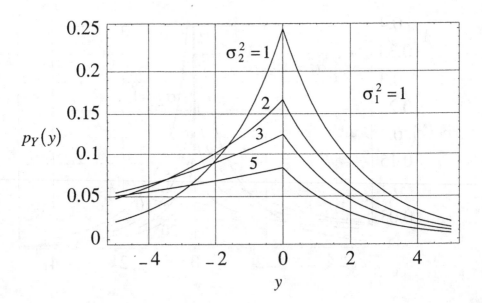

Fig. 13. cont'd.

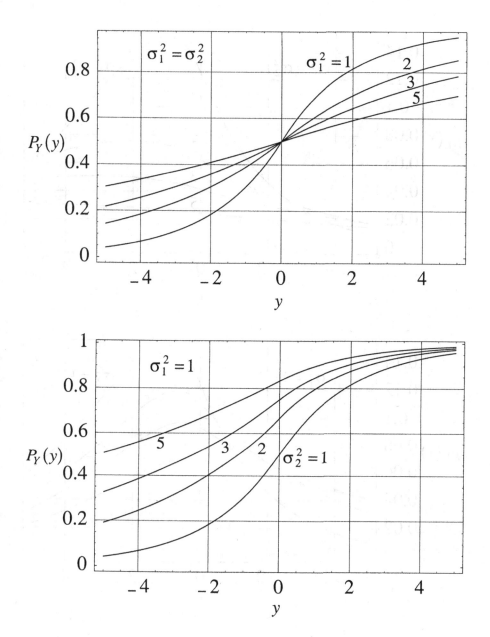

Fig. 14. Independent Central Chi-Square (-) Central Chi-Square CDFs: (a) $n_1=n_2=2$, equal variance, Eq. (4.5); (b) $n_1=n_2=2$, unequal variance, Eq. (4.5).

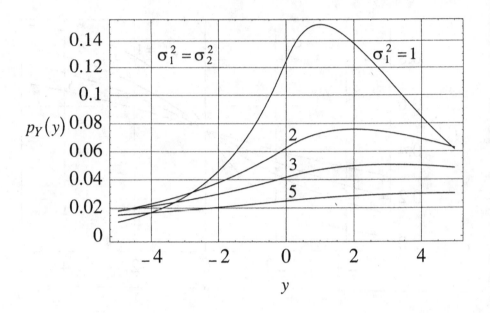

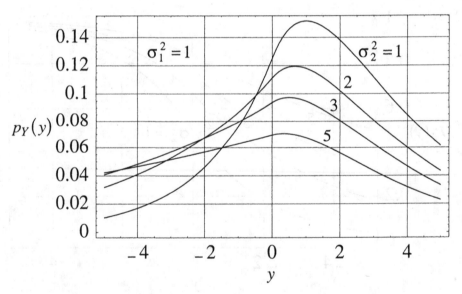

Fig. 15. Independent Central Chi-Square (-) Central Chi-Square PDFs: (a) $n_1=4$, $n_2=2$, equal variance, Eq. (4.13); (b) $n_1=4$, $n_2=2$, unequal variance, Eq. (4.13).

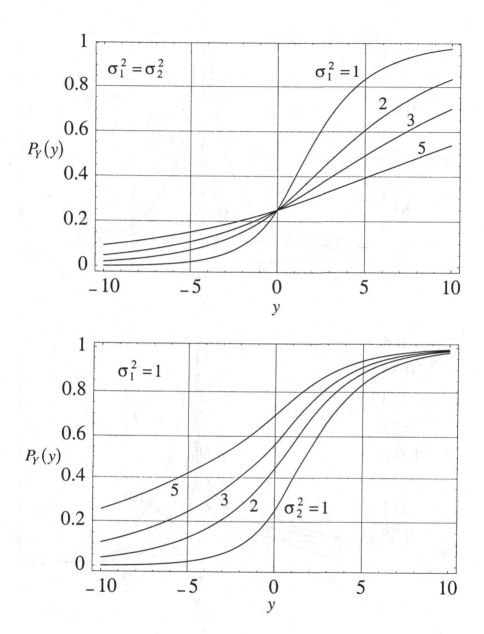

Fig. 16. Independent Central Chi-Square (-) Central Chi-Square CDFs: (a) $n_1=4$, $n_2=2$, equal variance, Eq. (4.14); (b) $n_1=4$, $n_2=2$, unequal variance, Eq. (4.14).

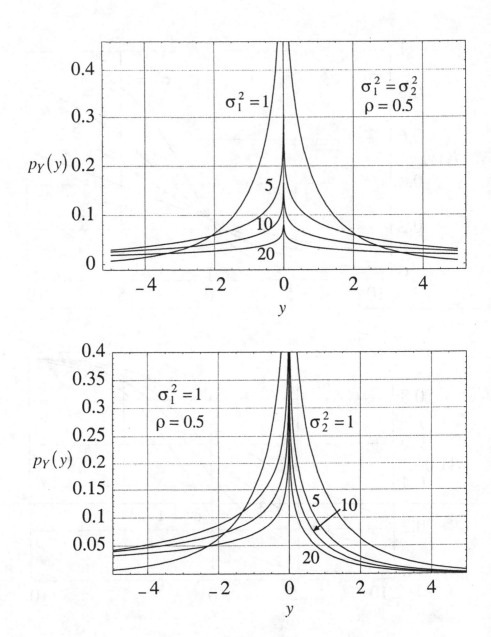

Fig. 17. Dependent Central Chi-Square (-) Central Chi-Square PDFs: (a) $n_1=n_2=1$, equal variance, Eq. (4.20); (b) $n_1=n_2=1$, unequal variance, Eq. (4.20); (c) $n_1=n_2=2$, equal variance, Eq. (4.23); (d) $n_1=n_2=2$, unequal variance, Eq. (4.23).

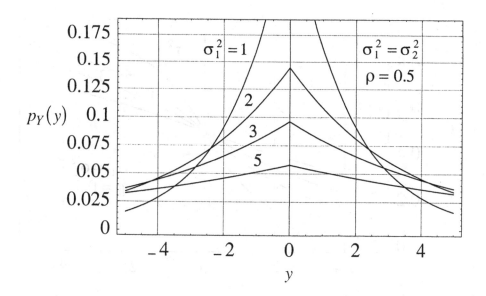

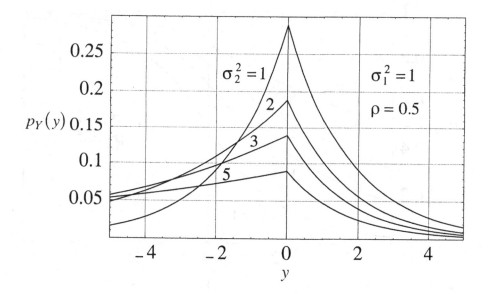

Fig. 17. cont'd.

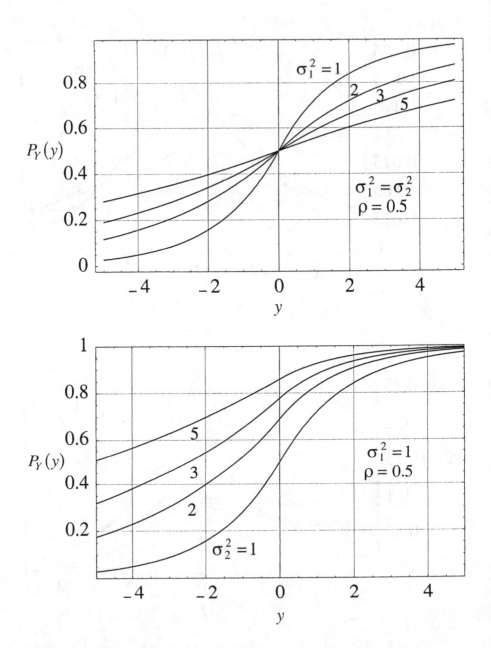

Fig. 18. Dependent Central Chi-Square (-) Central Chi-Square CDFs:
(a) $n_1 = n_2 = 2$, equal variance, Eq. (4.24); (b) $n_1 = n_2 = 2$, unequal variance,
Eq. (4.24).

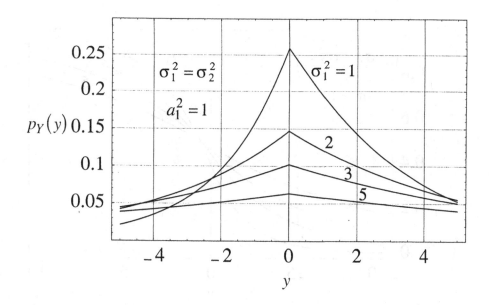

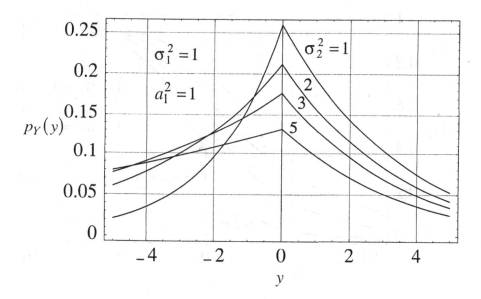

Fig. 19. Independent Noncentral Chi-Square (-) Central Chi-Square PDFs: (a) $n_1=n_2=2$, equal variance, Eq. (4.32); (b) $n_1=n_2=2$, unequal variance, Eq. (4.32).

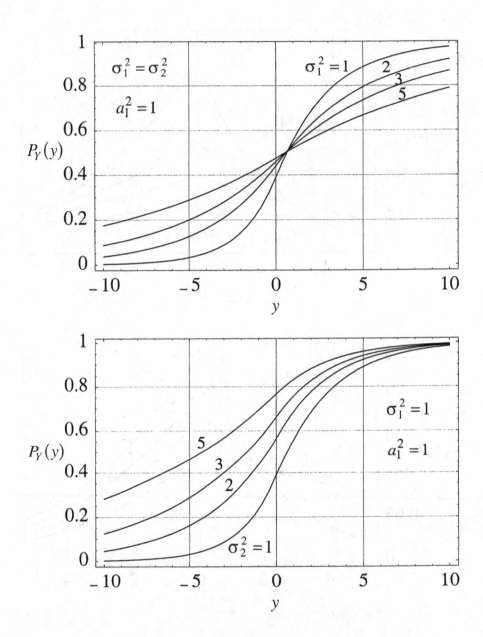

Fig. 20. Independent Noncentral Chi-Square (-) Central Chi-Square CDFs: (a) $n_1=n_2=2$, equal variance, Eq. (4.33); (b) $n_1=n_2=2$, unequal variance, Eq. (4.33).

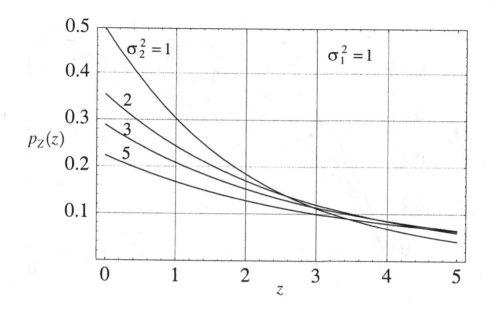

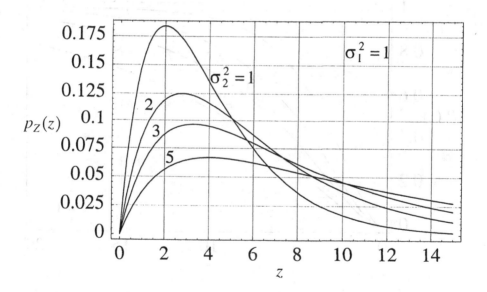

Fig. 21. Independent Central Chi-Square (+) Central Chi-Square PDFs: (a) $n_1=n_2=1$, unequal variance, Eq. (5.7); (b) $n_1=n_2=2$, unequal variance, Eq. (5.11).

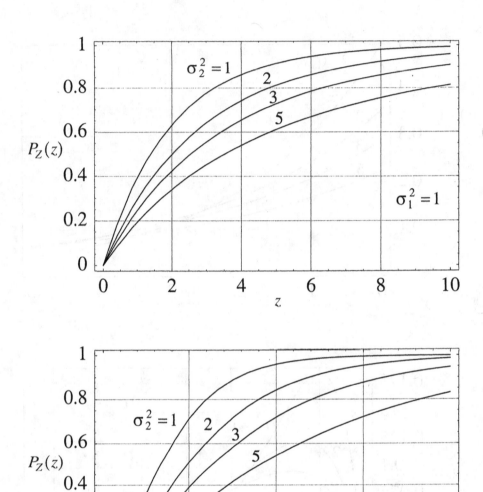

Fig. 22. Independent Central Chi-Square (+) Central Chi-Square CDFs: (a) $n_1=n_2=1$, unequal variance, Eq. (5.8); (b) $n_1=n_2=2$, unequal variance, Eq. (5.12).

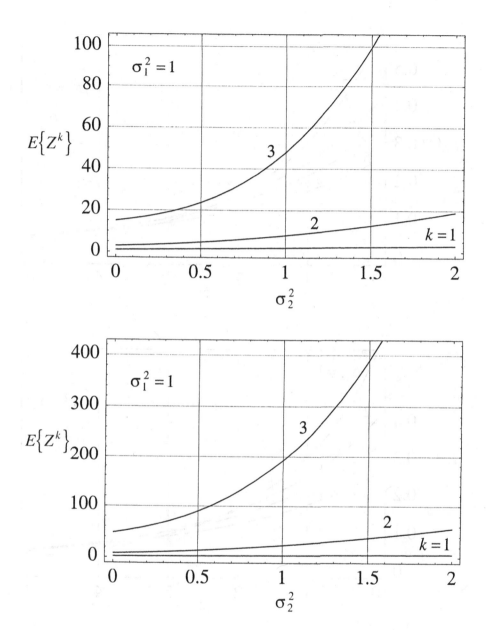

Fig. 23. Independent Central Chi-Square (+) Central Chi-Square Moments: (a) $n_1=n_2=1$, unequal variance, Eq. (5.10); (b) $n_1=n_2=2$, unequal variance, Eq. (5.14).

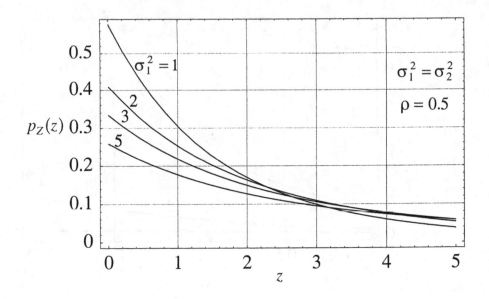

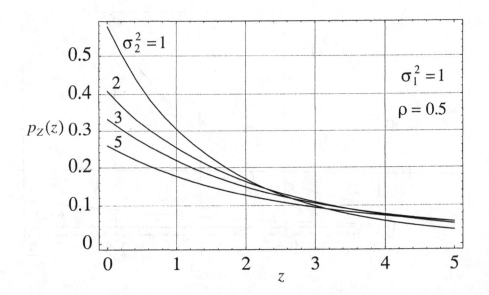

Fig. 24. Dependent Central Chi-Square (+) Central Chi-Square PDFs: (a) $n_1=n_2=1$, equal variance, Eq. (5.30); (b) $n_1=n_2=1$, unequal variance, Eq. (5.30); (c) $n_1=n_2=2$, equal variance, Eq. (5.34); (d) $n_1=n_2=2$, unequal variance, Eq. (5.34).

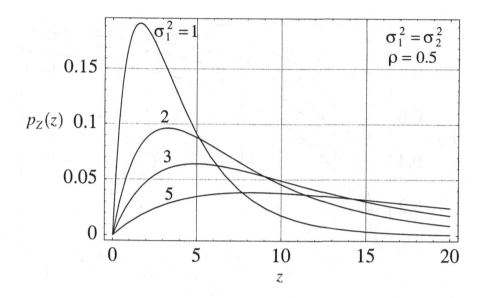

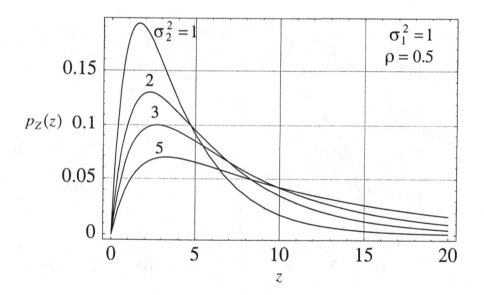

Fig. 24. cont'd.

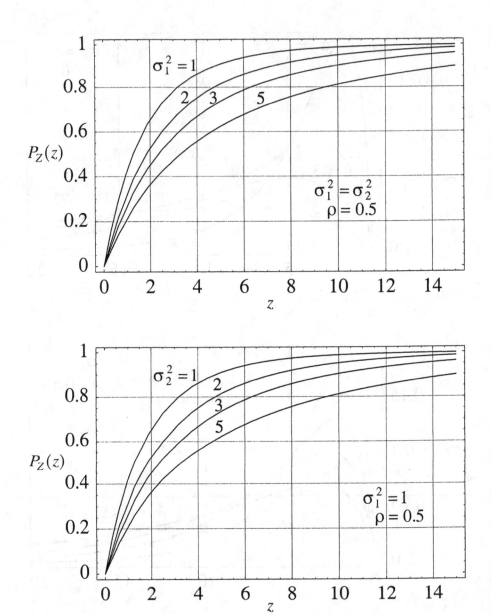

Fig. 25. Dependent Central Chi-Square (+) Central Chi-Square CDFs:
(a) $n_1=n_2=1$, equal variance, Eq. (5.31); (b) $n_1=n_2=1$, unequal variance,
Eq. (5.31); (c) $n_1=n_2=2$, equal variance, Eq. (5.35); (d) $n_1=n_2=2$, unequal
variance, Eq. (5.35).

172

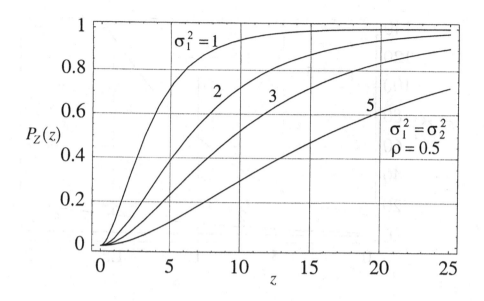

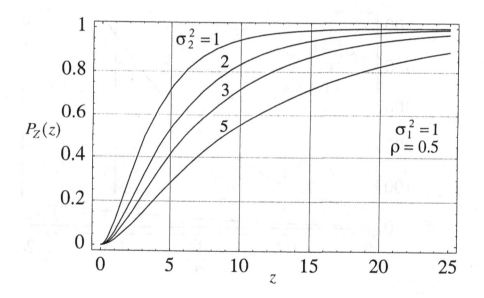

Fig. 25. cont'd.

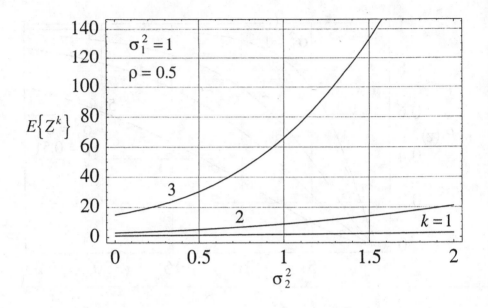

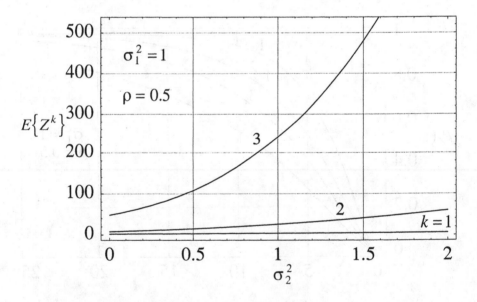

Fig. 26. Dependent Central Chi-Square (+) Central Chi-Square Moments: (a) $n_1=n_2=1$, Eq. (5.33); (b) $n_1=n_2=2$, Eq. (5.37).

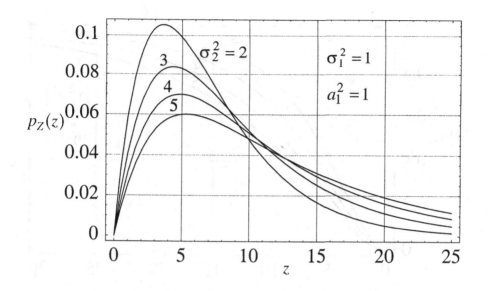

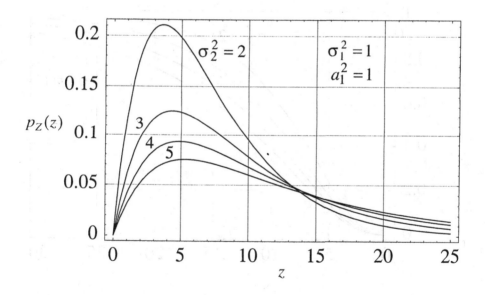

Fig. 27. Independent Noncentral Chi-Square (+) Central Chi-Square PDFs: (a) $n_1=n_2=2$, Eq. (5.45); (b) $n_1=4$, $n_2=2$, Eq. (5.48).

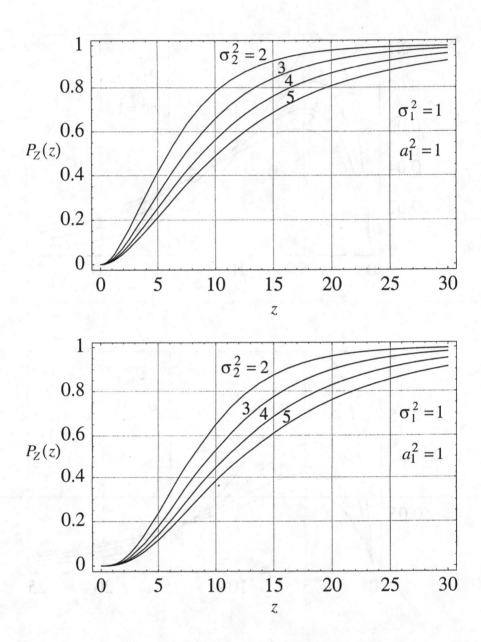

Fig. 28. Independent Noncentral Chi-Square (+) Central Chi-Square CDFs: (a) $n_1=n_2=2$, Eq. (5.46); (b) $n_1=4$, $n_2=2$, Eq. (5.49).

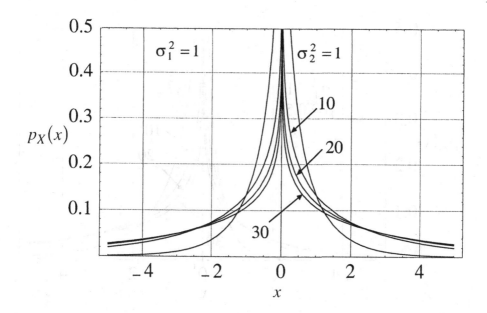

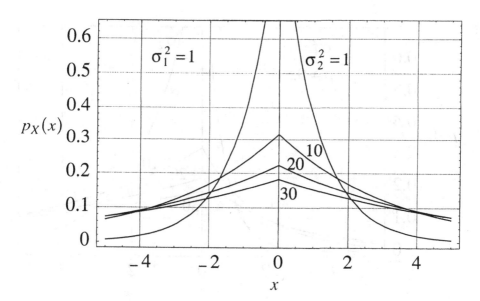

Fig. 29. Independent Zero Mean Gaussian (×) Gaussian PDFs:
(a) $n=1$, Eq. (6.2); (b) $n=2$, Eq. (6.5).

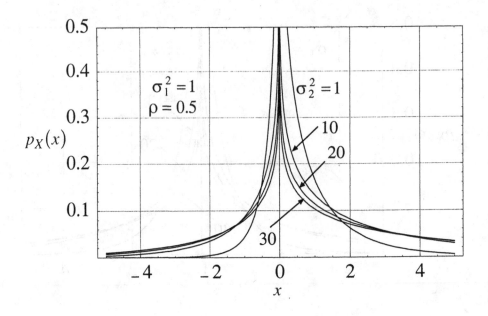

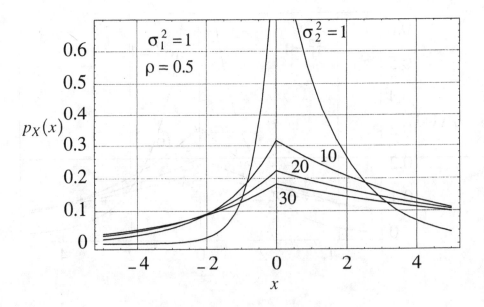

Fig. 30. Dependent Zero Mean Gaussian (×) Gaussian PDFs:
(a) $n=1$, Eq. (6.15); (b) $n=2$, Eq. (6.18).

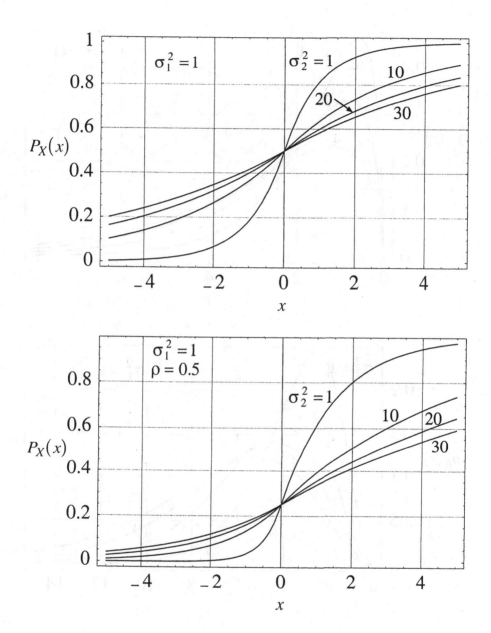

Fig. 31. Zero Mean Gaussian (×) Gaussian CDFs: (a) Independent, $n=2$, Eq. (6.6); (b) Dependent, $n=2$, Eq. (6.19).

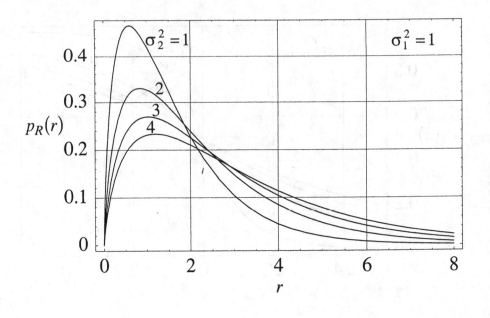

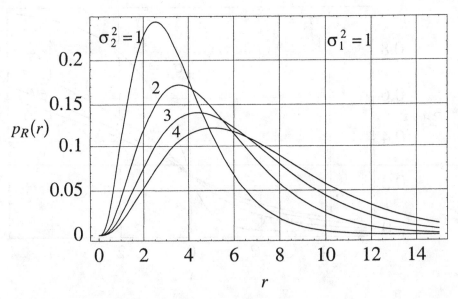

Fig. 32. Independent Rayleigh (×) Rayleigh PDFs: (a) $n_1=n_2=2$, Eq. (6.45); (b) $n_1=n_2=4$, Eq. (6.48).

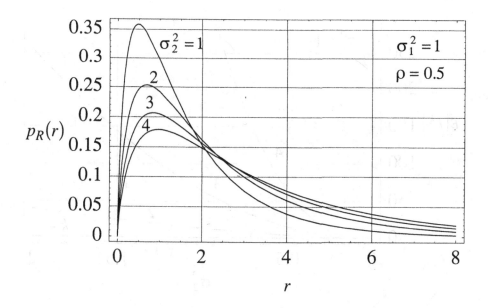

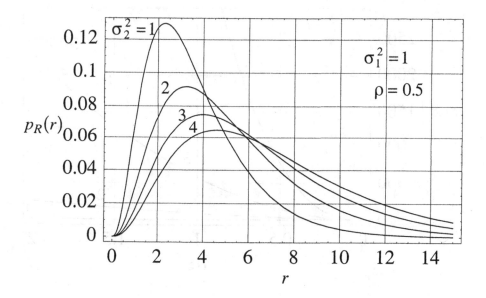

Fig. 33. Dependent Rayleigh ($\times$) Rayleigh PDFs: (a) $n=2$, Eq. (6.54); (b) $n=4$, Eq. (6.54).

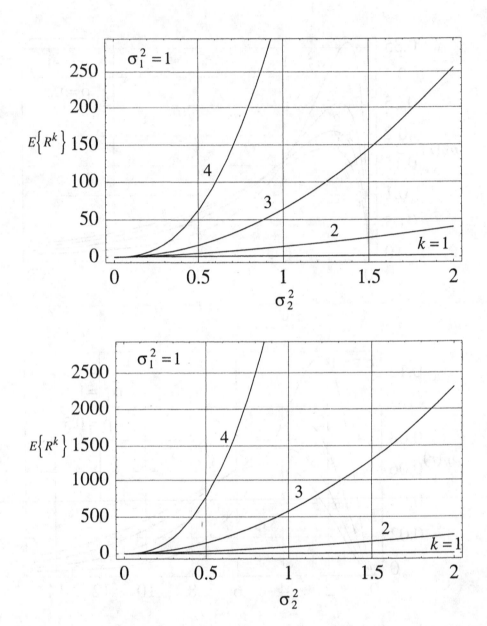

Fig. 34. Independent Rayleigh (×) Rayleigh Moments: (a) $n_1=n_2=2$, Eq. (6.47); (b) $n_1=n_2=4$, Eq. (6.50).

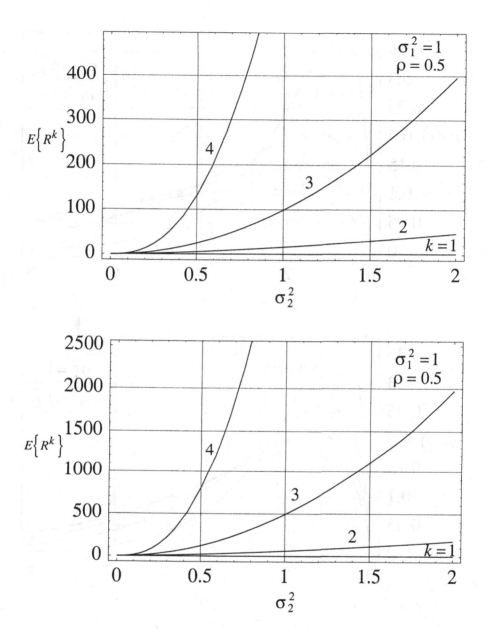

Fig. 35. Dependent Rayleigh (×) Rayleigh Moments: (a) $n=2$, Eq. (6.56);
(b) $n=4$, Eq. (6.56).

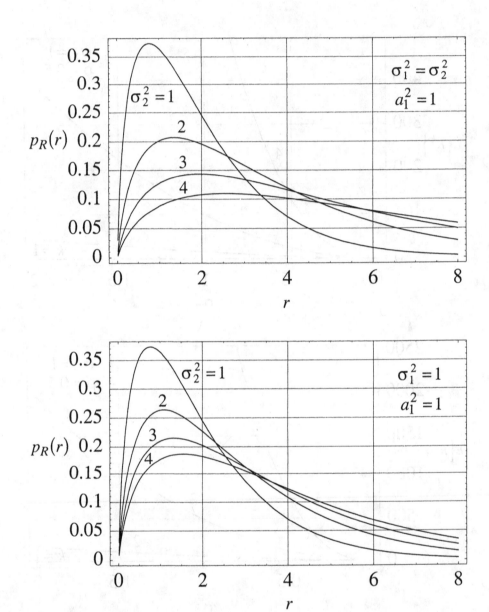

Fig. 36. Independent Rice ($\times$) Rayleigh PDFs: (a) $n_1=n_2=2$, equal variance, Eq. (6.59); (b) $n_1=n_2=2$, unequal variance, Eq. (6.59).

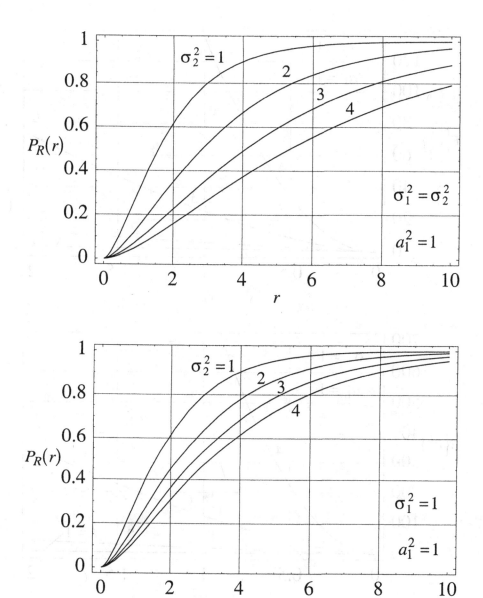

Fig. 37. Independent Rice (×) Rayleigh CDFs: (a) $n_1=n_2=2$, equal variance, Eq. (6.60); (b) $n_1=n_2=2$, unequal variance, Eq. (6.60).

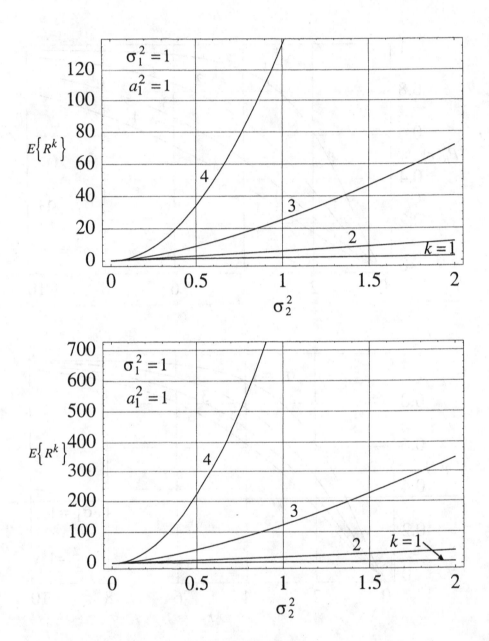

Fig. 38. Independent Rice (×) Rayleigh Moments: (a) $n_1=n_2=2$, Eq. (6.61); (b) $n_1=n_2=4$, Eq. (6.64).

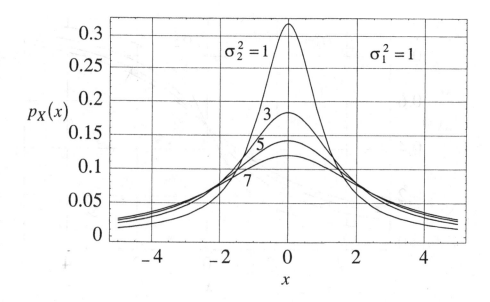

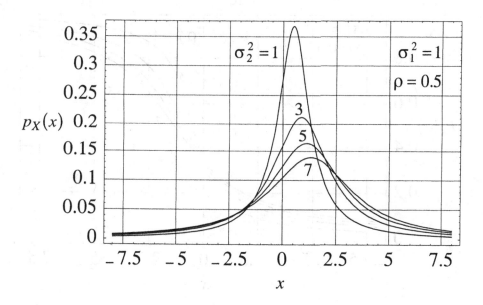

Fig. 39. Zero Mean Gaussian (÷) Gaussian PDFs: (a) Independent, Eq. (7.1); (b) Dependent, Eq. (7.9).

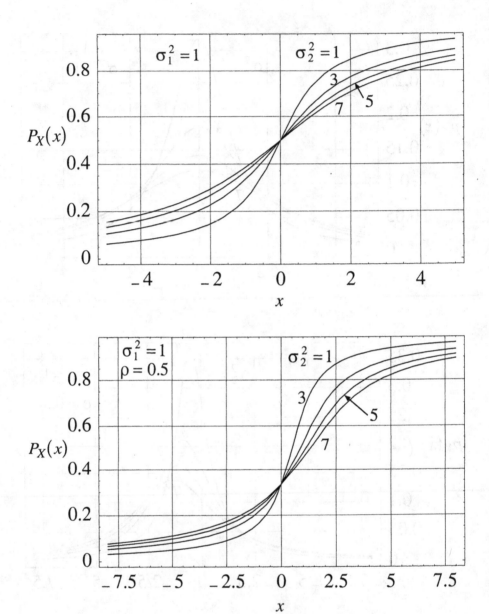

Fig. 40. Zero Mean Gaussian (÷) Gaussian CDFs: (a) Independent, Eq. (7.2); (b) Dependent, Eq. (7.10).

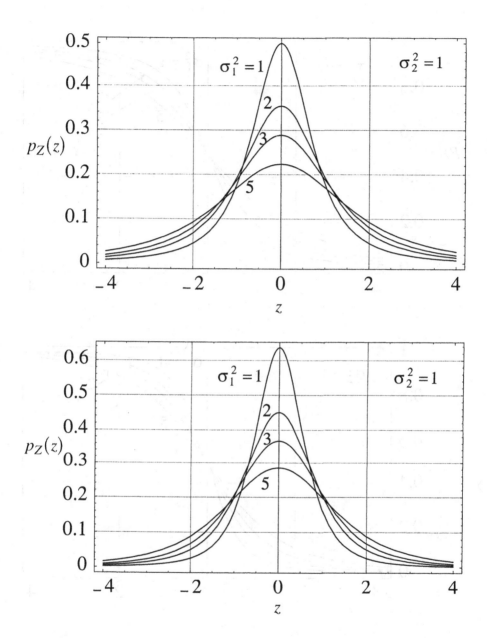

Fig. 41. Independent Zero Mean Gaussian ($\div$) Rayleigh PDFs:
(a) $n=2$, Eq. (7.19); (b) $n=3$, Eq. (7.21).

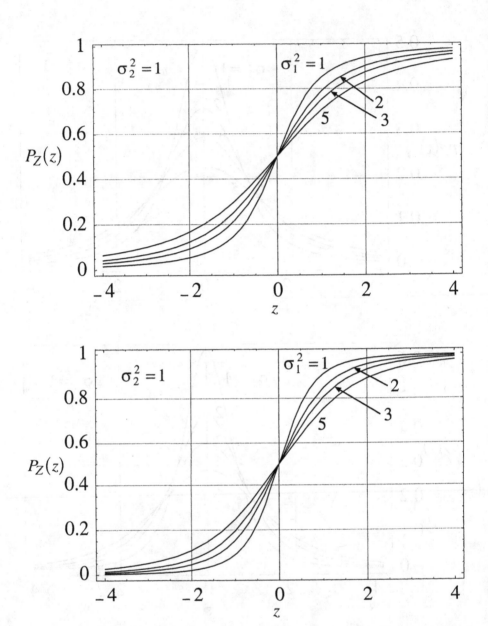

Fig. 42. Independent Zero Mean Gaussian ($\div$) Rayleigh CDFs: (a) $n=2$, Eq. (7.20); (b) $n=3$, Eq. (7.22).

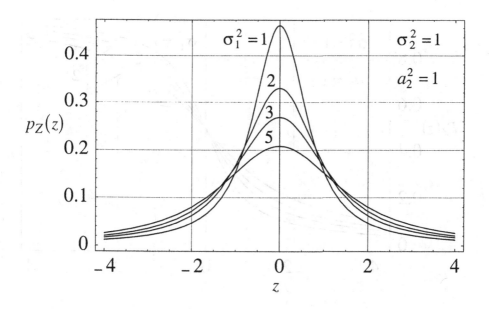

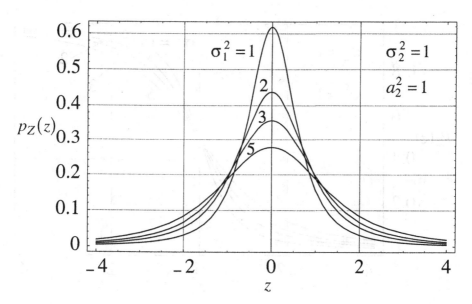

Fig. 43. Independent Zero Mean Gaussian ($\div$) Rice PDFs:
(a) $n=1$, Eq. (7.28); (b) $n=2$, Eq. (7.30).

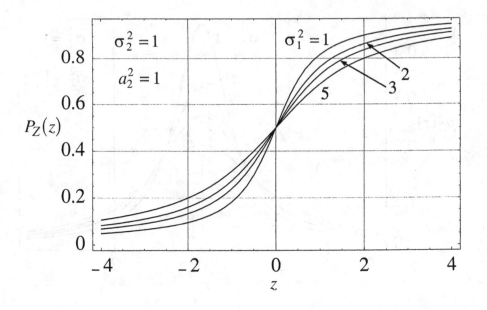

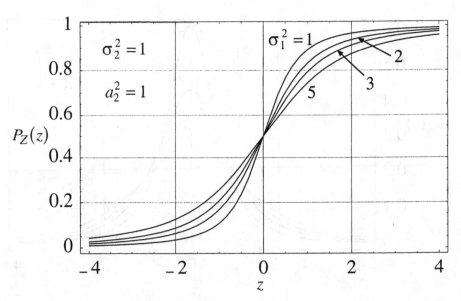

Fig. 44. Independent Zero Mean Gaussian (÷) Rice CDFs: (a) $n=1$, Eq. (7.29); (b) $n=2$, Eq. (7.31).

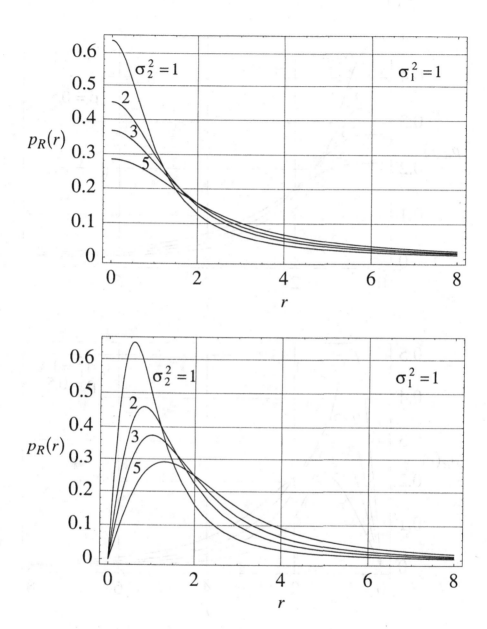

Fig. 45. Independent Rayleigh ($\div$) Rayleigh PDFs: (a) $n_1=n_2=1$, Eq. (7.37); (b) $n_1=n_2=2$, Eq. (7.44).

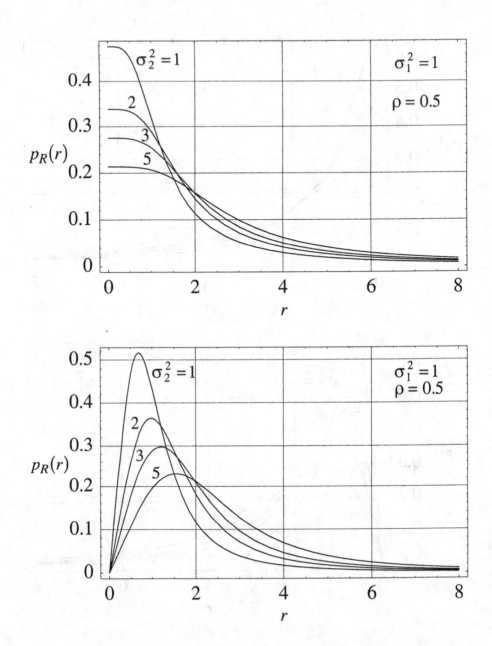

Fig. 46. Dependent Rayleigh (÷) Rayleigh PDFs: (a) $n=1$, Eq. (7.56); (b) $n=2$, Eq. (7.58).

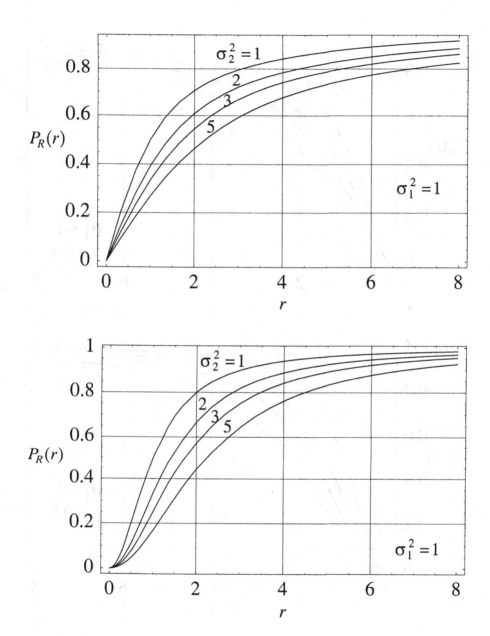

Fig. 47. Independent Rayleigh ($\div$) Rayleigh CDFs: (a) $n_1=n_2=1$, Eq. (7.38); (b) $n_1=n_2=2$, Eq. (7.45).

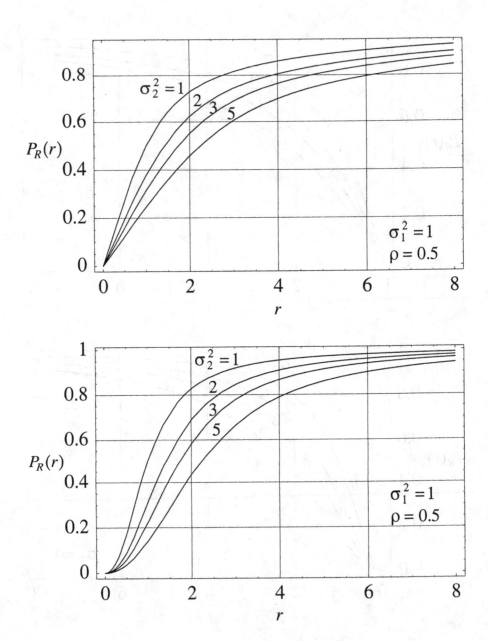

Fig. 48. Dependent Rayleigh (÷) Rayleigh CDFs: (a) $n=1$, Eq. (7.57); (b) $n=2$, Eq. (7.59).

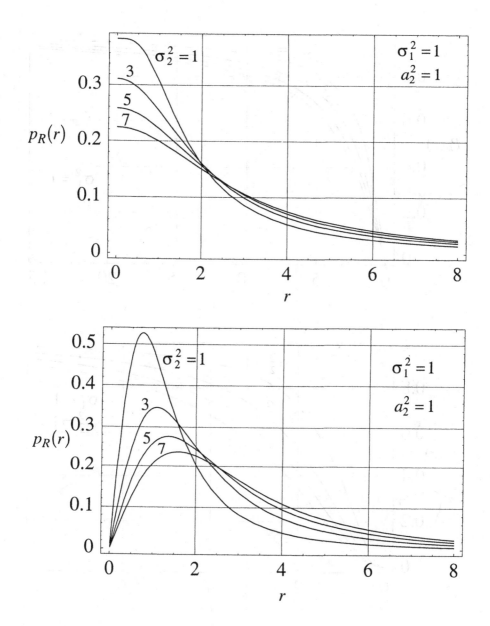

Fig. 49. Independent Rice ($\div$) Rayleigh PDFs: (a) $n_1=n_2=1$, Eq. (7.67); (b) $n_1=n_2=2$, Eq. (7.74).

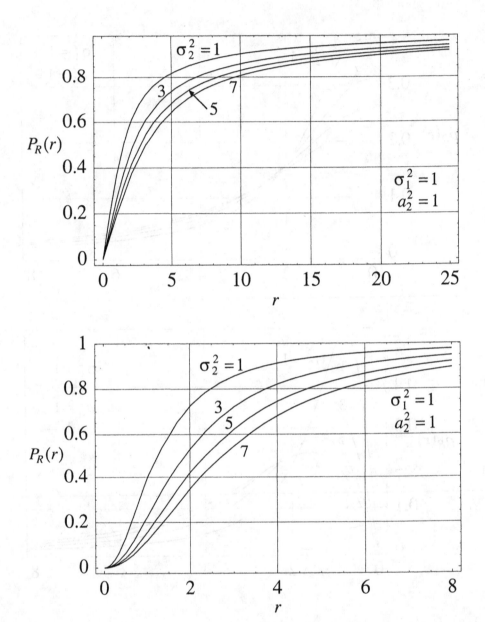

Fig. 50. Independent Rice (÷) Rayleigh CDFs: (a) $n_1=n_2=1$, Eq. (7.68); (b) $n_1=n_2=2$, Eq. (7.75).

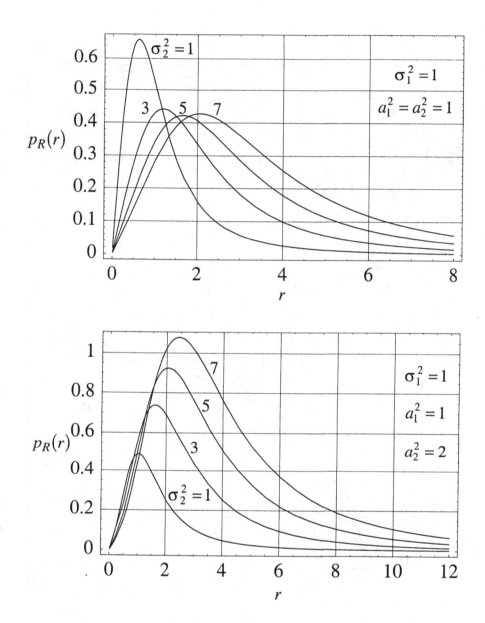

Fig. 51. Independent Rice ($\div$) Rice PDFs: (a) $n_1=n_2=2$, equal noncentrality parameters, Eq. (7.82); (b) $n_1=n_2=2$, unequal noncentrality parameters, Eq. (7.82).

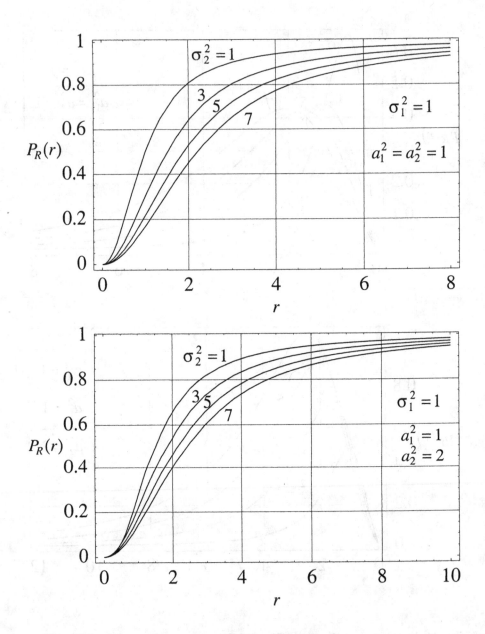

Fig. 52. Independent Rice (÷) Rice CDFs: (a) $n_1=n_2=2$, equal noncentrality parameters, Eq. (7.83); (b) $n_1=n_2=2$, unequal noncentrality parameters, Eq. (7.83).